T0135588

Model Based Signal Enhancement for Impulse Response Measurement

Von der Fakultät für Elektrotechnik und Informationstechnik der
Rheinischen-Westfälischen Technischen Hochschule Aachen
zur Erlangung des akademischen Grades eines

DOKTORS DER INGENIEURWISSENSCHAFTEN

genehmigte Dissertation

vorgelegt von

M.Sc.
Xun Wang
aus Sichuan (China)

Berichter:

Universitätsprofessor Dr. rer. nat. Michael Vorländer
Professor Dr.-Ing. Michael Möser

Tag der mündlichen Prüfung: 17.12 2013

Diese Dissertation ist auf den Internetseiten der Hochschulbibliothek online verfügbar.

Xun Wang

Model Based Signal Enhancement for Impulse Response Measurement

Logos Verlag Berlin GmbH

λογος

Aachener Beiträge zur Technischen Akustik

Editor:
Prof. Dr. rer. nat. Michael Vorländer
Institute of Technical Acoustics
RWTH Aachen University
52056 Aachen
www.akustik.rwth-aachen.de

Bibliographic information published by the Deutsche Nationalbibliothek

The Deutsche Nationalbibliothek lists this publication in the Deutsche Nationalbibliografie; detailed bibliographic data are available in the Internet at http://dnb.d-nb.de .

D 82 (Diss. RWTH Aachen University, 2013)

ISBN 978-3-8325-3630-5
ISSN 1866-3052
Vol. 18

Logos Verlag Berlin GmbH
Comeniushof, Gubener Str. 47,
D-10243 Berlin
Tel.: +49 (0)30 / 42 85 10 90
Fax: +49 (0)30 / 42 85 10 92
http://www.logos-verlag.de

Contents

Glossary

Acronyms

BSS	Blind Source Separation
FFT	Fast Fourier Transform
FHT	Fast Hadamard Transform
ICA	Independent Component Analysis
LSE	Least-Square Estimation
LTI	Linear Time-Invariant
MIMO	Multiple Input Multiple Output
MLS	Maximum Length Sequence
SNR	Signal-to-Noise Ratio
SUT	Strong Uncorrelating Transform
SVD	Singular Value Decomposition
TDS	Time Delay Spectrometry

Abstract

Impulse response measurements that are performed outdoors are highly susceptible to uncertainties caused by the non-perfect measurement setup, the presence of background noise, and fluctuations in media such as wind and temperature drift. This work concentrates on two scenarios: the measurement of reflection coefficients of noise barriers and the influence of temperature variances in machinery cavities.

Regarding the reflection coefficient measurement, an optimized microphone array is implemented to separate direct sound and reflected sound. Compared with the standard subtraction method, it is possible to obtain the reflection coefficient through only one single measurement without moving the devices between the free-field room and the sound barrier under test, and to avoid the errors resulting from an imperfect measurement setup and time variances throughout the procedure.

For the purpose of de-noising, the option of using statistics-based source separation methods is also studied. Simulation results show that source separation can indeed reduce the background noise effect in a reverberant environment. However, it cannot exceed the performance of synchronous averaging. When the excitation signal is known and no specific knowledge about the noise can be used, averaging is the most efficient way to improve the SNR for time-invariant systems.

The application of long-time averages, however, runs the risk of time variances. The possibility of phase-shift compensation in wind fluctuations is analysed here. The time-varying phase shift can be compensated for the direct sound component. The reflection coefficient measurement has more complex effects, and both the magnitude and the phase are changed. The influence of wind fluctuations in a field consisting of many reflections cannot be compensated for by using the same approach.

Temperature variances also influence the accuracy of impulse response measurements. Concerning an online machine monitoring scenario, the temperature drifts during the transfer function measurement and the speed of sound varies with temperature in the machine environment. As a consequence, the impulse response is stretched along the time axis. A time-warping model is derived and applied to compensate for inter-period (slow) and intra-period (rapid) temperature variances. In this way, measurements of higher accuracy can be obtained.

1 Introduction

The impulse response and the associated transfer function are the essential properties of linear acoustic systems. These properties must be measured with an appropriate accuracy. Accuracy is easily guaranteed under laboratory conditions because the temperature and other parameters causing time variances can be controlled rather well. However, if the measurement is performed in the field (in situ), the accuracy is highly susceptible to environmental variances, such as the presence of noise, wind fluctuation and temperature drifts. In case background noise is present, synchronous averaging is a straightforward method to reduce the noise influence, but the operator might run the risk of time variances during long-term averaging. For example, [Vorländer and Kob, 1997] examine the errors of measurement using Maximum Length Sequence (MLS), which are caused by wind speed fluctuations and temperature variances. [Svensson and Nielsen, 1999] quantify both inter-period and intra-period time variance, and show that both kinds of variances cause apparent energy-level losses in the transfer function. When sweeps are used as excitation signals, a long single sweep shows better immunity against time variances than the average of multiple short sweeps[Fumiaki Satoh, 2002] .

Besides synchronous averaging, some other signal processing techniques—such as spatial filtering, Blind Source Separation (BSS) and blind deconvolution—might be potential ways to extract noise from the desired signals. The advantage of spatial filtering is that the desired signal and the noise can even overlap in the time and frequency domains. It can separate the noise through the different directions of incoming waves. However, the spatial filter can separate only a limited number of waves (which might be also reflections). When the number of reflections increases, the computational cost becomes prohibitive. The advantage of BSS is that not much a-priori knowledge about the source is required. A huge number of statistical methods have been studied [Jutten, 2010][Haykin, 1994]. Most of these assume that precise knowledge of an input signal is inaccessible but that the statistical independence properties of the sources can be used. However, no all-conquering BSS algorithms are available to separate the sources and noises for all scenarios because statistical signal processing methods always have to trade off the unknown discriminative parameters and useful samples. The discriminative parameters differ from one scenario to another.

This dissertation considers two specific scenarios.

First, we consider the reflection coefficient measurements of sound barriers. Reflection and absorption coefficient measurements are the fundamental measurements for the acoustic properties of surfaces. The standard measurement setups are the Kundt's tube [ISO, 1998] [AST] and the reverberation chamber [ISO, 2003]. Sometimes, however, the measurement setup has to be performed *in situ*, and accordingly some other problems have to be tackled. For example, the sound barrier has been constructed near a highway, and the reflection coefficient or the absorption coefficient have to be measured at the site and in the presence of high background noise. In this case, reflections from the surface can be measured by using the impulse response technique. The direct sound and the reflection have to be separated and analysed. Direct sound and reflected sound are represented as two impulses in the overall impulse response of the system. If the direct sound and the reflection are ideal Dirac impulses, they can be simply windowed out in the time domain. In practice, however, they are not perfect impulses, are very close to each other, and even overlap. The subtraction procedure can be used [Mommertz, 1995], but it requires an additional free-field measurement as a reference. Another inverse filtering method is developed to equalize the frequency response of the loudspeaker, and so that the loudspeaker radiates shorter pulses [Cobo, 2007] [Wehr, 2013]. This dissertation studies an optimized microphone array that can be used to separate direct sound and reflection sound.

The *in situ* measurements are usually contaminated by surrounding noise. The possibilities of applying the statistical BSS method to separate the external noise source are investigated. In case many noise sources appear randomly—for example, vehicles passing the noise barrier successively during the measurement—no specific model can be applied to separate the noise. Synchronous averaging is the most efficient way to improve the Signal-to-Noise Ratio (SNR). If the synchronous average is performed over a long period, however, the influence of time variances cannot be ignored. There are two typical time variances: wind fluctuation and temperature shift. Wind fluctuation leads to phase shifts. The possibility of compensating for the effect of the time-varying phase shifts is discussed herein.

In the reflection coefficient measurement, the influence of temperature variance is not a severe problem. However, in the second scenario discussed in this dissertation—i.e. the machine diagnosis scenario—the temperature might drift by several degree Celsius during long-time measurement. A temperature shift changes the speed of sound, and the speed of sound influences the impulse response by a time-stretching process. In order to obtain higher measurement accuracy, this time-stretching process must be studied.

The dissertation is organized as follows. In Chapter 2, the fundamentals of the impulse response measurement, the spatial filtering and statistical signal processing are briefly described. Chapter 3 focuses on the reflection coefficient *in situ* measurement. The spatial filter-based method is discussed here. In Chapter 4, the possibility of noise separation with statistical signal processing is investigated. Chapter 5 concerns the wind fluctuations affecting reflection coefficient measurement, and the possibility to compensate for the phase shift is discussed. In Chapter 6, a time-warping model for compensation of the temperature shifts is investigated.

2 Fundamentals

2.1. Basics in signal processing

The system determined by the differential equations can be expressed by the mathematical framework of Linear Time-Invariant (LTI). The relation between the input and output is formulated by the convolution theory as [Alan Victor Oppenheim, 1997]

$$
\begin{aligned}
x(t) &= \int_{-\infty}^{\infty} x(t)h(t-\tau)\mathrm{d}t \\
&= h(t) * s(t)
\end{aligned}
\tag{2.1}
$$

where $s(t)$ is the input and $x(t)$ is the output of the system. The relation between the input and the output is the convolution with the system response $h(t)$, named as the **Impulse Response**.

Using the discrete-time notation, the convolution is written as Eq. 2.2

$$
\begin{aligned}
x(n) &= \sum_{k=-\infty}^{\infty} x(k)h(n-k) \\
&= h(n) * s(n)
\end{aligned}
\tag{2.2}
$$

The signals can also be represented in the frequency domain by Fourier transform. The definition of Fourier transform is

$$X(\omega) = \int_{-\infty}^{\infty} x(t)e^{-i\omega t}\mathrm{d}t$$
$$X(t) = \frac{1}{2\Pi}\int_{-\infty}^{\infty} x(\omega)e^{i\omega t}\mathrm{d}\omega \tag{2.3}$$

where $X(\omega)$ is called the complex spectrum of the signal $x(t)$, and the Fourier transform can be nowadays computed very efficiently by Fast Fourier Transform (FFT). One essential property of the Fourier transform is the spectrum of the convolution of two signals is the multiplication of the two signals in frequency domain, as Eq. 2.4

$$s(t) = h(t) * s(t) \xLeftrightarrow{\text{Fourier transform}} X(\omega) = H(\omega)S(\omega) \tag{2.4}$$

In this way, the convolution can be efficiently computed by multiplying the two signals in frequency domain and transforming the overall spectrum back to the time domain as Eq. 2.5.

$$x(t) = \frac{1}{2\pi}\int_{-\infty}^{\infty} H(\omega)S(\omega)e^{i\omega t}\mathrm{d}t \tag{2.5}$$

In this dissertation, the term **Transfer Function** refers particularly to $H(\omega)$, which is complex spectrum of the impulse response $h(t)$. Most of the physical properties can be obtained by analysing the relative value at different frequencies, and the absolute physical quantity—sound pressure—can be obtained just by multiplying a calibration factor. Focusing on the signal processing technique, all the figures of transfer function are plotted just by the spectrum $H(\omega)$ in this dissertation, and the calibration factor is ignored.

2.2. Impulse response of acoustic systems

The acoustic properties, such as the reflections, delays, resonance frequencies, etc, can be expressed by the mathematical framework of impulse responses.

Theoretically, the impulse response can be derived through Green's function theory or the wave field theory. Eq. 2.6 is the homogeneous wave equation. The velocity potential is used here only for the sake of mathematical convenience. From the velocity potential, we can derive the most interesting quantities in acoustics: sound pressure by $p = \rho_0 \frac{\partial \Psi}{\partial t}$.

$$\begin{cases} \nabla^2 \Psi - \dfrac{1}{c_0^2} \dfrac{\partial^2 \Psi}{\partial t^2} = 0 \\ c_0 = \sqrt{\dfrac{\gamma R T_0}{M}} \\ p = \rho_0 \dfrac{\partial \Psi}{\partial t} \\ \vec{v} = -\nabla \Psi \end{cases} \tag{2.6}$$

The corresponding boundary condition is

$$\rho_0 \frac{\partial \Psi}{\partial t} + (\vec{n} \cdot \nabla \Psi) Z \Big|_{\varphi(\vec{r})} = 0 \tag{2.7}$$

where the meanings of the symbols are

Ψ	velocity potential
T_0	absolute temperature
γ	adiabatic constant
R	ideal gas constant
M	Molar mass
p	sound pressure
$\vec{v}$	particle velocity
ρ_0	density of air
Z	the boundary's impedance
$\varphi(\vec{r})$	the boundary surface
$\vec{n}$	the unit normal vector to the boundary surface

Assuming that one point source is located at $r = r_0$ and radiates the sound of the harmonic frequency $e^{i\omega t}$, the wave equation is re-written as

$$\nabla^2\Psi - \frac{1}{c_0^2}\frac{\partial^2\Psi}{\partial t^2} = \delta(\vec{r} - \vec{r}_0)e^{i\omega t} \tag{2.8}$$

The steady state solution is notated as $\Psi = \Psi(\vec{r}, \omega)e^{i\omega t}$, where

$$\nabla^2\Psi(\vec{r}, \omega) + (\frac{\omega}{c_0})^2 \cdot \Psi(\vec{r}, \omega) = \delta(\vec{r} - \vec{r}_0) \tag{2.9}$$

The solution of Eq. 2.9 can be expressed by the Helmholtz-Huygens integral as Eq. 2.10 [Philip M. Morse, 1968].

$$\begin{aligned}\Psi(\vec{r}, \omega) = & \iiint_\Omega Q(\vec{r}_0)G(\vec{r}, \vec{r}_0, \omega)\mathrm{d}\vec{r}_0 \\ & - \iint_S \left[G(\vec{r}, \vec{r'}, \omega)\frac{\partial\Psi(\vec{r'}, \omega)}{\partial n} - \Psi(\vec{r'}, \omega)\frac{\partial G(\vec{r}, \vec{r'}, \omega)}{\partial n}\right]\mathrm{d}S\end{aligned} \tag{2.10}$$

where $\Psi(\vec{r}, \omega)$ denotes the velocity potential for frequency ω at the location point $\vec{r}$, and $Q(\vec{r}_0)$ is the source function $\big(Q(\vec{r}_0) = \delta(\vec{r} - \vec{r}_0)\big)$ for Eq. 2.9). The first integration given in Eq. 2.10, $\iiint_\Omega Q(\vec{r}_0)G(\vec{r}, \vec{r}_0, \omega)\mathrm{d}\vec{r}_0$, stands for the direct sound from the source. The second integration is the contribution from the reflection caused by the boundaries. $\vec{r'}$ stands for the boundary's coordinate.

$G(\vec{r}, \vec{r'}, \omega)$ and $G(\vec{r}, \vec{r}_0, \omega)$are Green's functions for a point source in free space (Eq. 2.11), which is the solution of the inhomogeneous Helmholtz equation in free space, Eq. 2.12.

$$G(\vec{r}, \vec{r}_0, \omega) = \frac{e^{-ikr}}{4\pi r} = \frac{e^{-ik\sqrt{(x-x_0)^2+(y-y_0)^2+(z-z_0)^2}}}{4\pi\sqrt{(x-x_0)^2+(y-y_0)^2+(z-z_0)^2}} \tag{2.11}$$

where $k = \frac{\omega}{c_0}$.

$$\nabla^2 G(\vec{r}, \vec{r}_0, \omega) + \frac{\omega^2}{c_0^2}G(\vec{r}, \vec{r}_0, \omega) = -\delta(\vec{r} - \vec{r}_0) \tag{2.12}$$

In a bounded environment, the solution can also be expressed by the eigenfunction theory [Mikio Tohyama, 1998] [Tohyama, 2011], as shown in Eq. 2.13 as

$$\Psi(\vec{r},\omega) = \sum_n \frac{\Psi_n(\vec{r_0})\Psi_n^*(\vec{r})}{k_n^2 - \left(\frac{\omega}{c_0}\right)^2} \tag{2.13}$$

where $\Psi_n(\vec{r})$ is the orthogonal eigenfunction determined by the boundary condition, and k_n is the eigenvalue of the wave number.

The solution Eq. 2.10 and Eq. 2.13 , notated as $H(\vec{r},\vec{r_0},\omega)$, is the physical property of the acoustic system and unrelated to the excitation signal. For a given excitation $s(t)$, the wave equation is written as Eq. 2.14

$$\nabla^2\Psi - \frac{1}{c_0^2}\frac{\partial^2\Psi}{\partial t^2} = \delta(\vec{r}-\vec{r_0})s(t) \tag{2.14}$$

The solution is

$$\Psi(\vec{r},t) = \int_{-\infty}^{\infty} H(\vec{r},\vec{r_0},\omega)S(\omega)e^{i\omega t}\mathrm{d}\omega \tag{2.15}$$

where the $H(\vec{r},\vec{r_0},\omega)$ is the solution of time-independent Eq. 2.9, $S(\omega)$ is the spectrum of the excitation. Notating $h(\vec{r},\vec{r_0},t)$ as the inverse Fourier transform of $H(\vec{r},\vec{r_0},\omega)$, Eq. 2.15 tells that the sound field recorded at position $\vec{r}$ is actually the convolution of the excitation $s(t)$ with the impulse response $h(\vec{r},\vec{r_0},t)$.

Then $H(\vec{r},\vec{r_0},\omega)$ is accordingly defined as the transfer function from position $\vec{r_0}$ to position $\vec{r}$, and $h(\vec{r},\vec{r_0},t)$ can be defined as the impulse response as well. The relation between the wave equation and impulse response is thoroughly explained in the book [Mikio Tohyama, 1998] [Tohyama, 2011]. Since the analytical solution for the above transfer function equations exists only for a simple boundary condition, the boundary element method or finite element method is typically applied to calculate the impulse response numerically.

The impulse response is determined by the wave equation and is relative to the speed of sound and the movements of the medium (usually air). In general, the speed of sound depends on the temperature. If the temperature and the medium change during the measurement, the impulse response will change consequently. The possibility that compensates for the effects of such variations in time-variant systems will be described in this thesis.

2.3. Impulse response measurement with various excitation signals

When the sound is radiated to a receiver position, the sound at the receiver's position is the convolution of the sound source signal with the impulse response, as shown by Eq.2.16. $s(t)$ is the source signal and $x(t)$ is the measured signal. $h(t)$ is the impulse response of the acoustic system.

$$x(t) = h(t) * s(t) \tag{2.16}$$

The relation between the excitation signal and the measured signal can be represented in the frequency domain, as in Eq.2.17.

$$X(\omega) = H(\omega)S(\omega) \tag{2.17}$$

Theoretically, in order to measure the impulse response of an acoustic system, the excitation signal can be any kind of broadband signal. As long as the excitation signal is known, the impulse response can be directly calculated by denvolving the measured signal with the excitation signal.

Before the digital computers were introduced as part of system measurement methods, other techniques were used to obtain transfer functions. The level recorder is one of the oldest [E. C. Wente, 1935]. A swept-sine signal is generated through an analogue generator, the resulting voltage is input to an differential amplifier and then to a potentiometer. The output of the potentiometer is linked to a writing pen. Finally, the voltage of the frequency response is written on a sheet of coordinate paper. This technique requires no digital circuitry. The Time Delay Spectrometry (TDS) is another approach to measure the transfer function based on analogue devices. Compared with the level recorder, TDS is capable of measuring both, the amplitude and the phase response. It was introduced by Heyser [Heyser, 1967],[Heyser, 1969a],[Heyser, 1969b] especially for the measurement of loudspeakers. It can also be applied to any other linear acoustic measurements.

Finally, with the help of digital circuitry and advanced signal processing algorithms, the transfer function can be measured accurately. This technique entered the field of acoustics in the 1970s. [Berman, 1977] applied the computer to store and average the impulse response for loudspeaker measurement. The FFT is applied by [Berman, 1977] as well, but the recorded length of the impulse response is only a few milliseconds. The Legendre sequence, which is a pseudo-random noise sequence, is introduced by [Schroeder, 1979] to measure the room impulse responses. The advantage of the Legendre sequence is that its discrete Fourier transform is equal to (within a constant factor) the sequence itself, which simplifies its digital representation for storage and transmission. The MLS became popular in 1980s [Borish, 1983] [Borish, 1985] [Rife, 1989]. MLS is also a pseudo-random noise sequence that can carry enough energy and cover the broadband frequency range. Since the auto-correlation function of such a pseudo-random sequence is the Dirac function, the impulse response can be recovered by the cross-correlation of the pseudo-random sequence with the recorded signal (as Eq.2.18). The cross-correlating process is computed by Fast Hadamard Transform (FHT)[Borish, 1983] [Borish, 1985]. FHT uses only addition and subtraction to calculate the cross-correlation, which can operate faster than the FFT of the same block length. This advantage of MLS is very useful when the memory size and computing power are limited two decades ago. However, this computing efficiency is no longer of particular interest since the processing time dedicated to the FHT and FFT are negligible nowadays.

$$x(t) * s(-t) = h(t) * s(t) * s(-t) = h(t) * \delta(t) = h(t) \tag{2.18}$$

In contrast, MLS shows very weak tolerance against time variance [Vorländer and Kob, 1997][Svensson and Nielsen, 1999] and the harmonic distortion of loudspeakers [Müller and Massarani, 2001].

Since the year 2000, sweep have become the preferred excitation signals because of their superior ability to handle time variances and distortions [Farina, 2000] [Farina, 2007b] [Farina, 2007a] [Satoh, 2004] [Fumiaki Satoh, 2004]. Sweep can be basically designed in two different ways: linear sweeps and logarithmic sweeps (also called exponential sweeps).

The analytical expression for a linear sweep is Eq.2.19[Farina, 2000].

$$s(t) = \sin\left(\omega_1 t + \frac{\omega_2 - \omega_1}{T} \cdot \frac{t^2}{2}\right) \tag{2.19}$$

where ω_1 is the start frequency, ω_2 is the stop frequency, and T is the length of the signal. The group delay or the corresponding instantaneous frequencies is increasing linearly as Eq. 2.20.

$$\omega(t) = \omega_1 + \frac{\omega_2 - \omega_1}{T} \cdot t \tag{2.20}$$

The phase of sweep can also be adjusted as exponentially increasing, as shown in Eq. 2.21. The start frequency of logarithmic sweep cannot be set to zero because of the term $\log \frac{\omega_2}{\omega_1}$, instead the start frequency must be set a little bit higher.

$$s(t) = \sin\left[\frac{\omega_1 \cdot T}{\log\left(\frac{\omega_2}{\omega_1}\right)} \cdot e^{\frac{t}{T}\log\left(\frac{\omega_2}{\omega_1}\right)}\right] \tag{2.21}$$

The instantaneous frequency of the logarithmic sweep is

$$\omega(t) = \omega_1 \cdot \left(\frac{\omega_2}{\omega_1}\right)^{\frac{t}{T}} \tag{2.22}$$

When the signal is recorded, the transfer function is obtained by dividing the spectrum of recorded signal by the spectrum of the sweep. The impulse response is calculated by the inverse Fourier transform of the transfer function.

$$H(\omega) = \frac{X(\omega)}{S(\omega)} \tag{2.23}$$

If the sweep does not cover the full frequency band, the equation is divided by zeros at the frequencies where the sweep does not cover. The regularization methods have to be used [Farina, 2007b] [Farina, 2007a] to avoid dividing by zeros.

Sweeps are superior to the MLS measurement. Firstly, sweeps offer a higher tolerance for time variance. Both the group delays of the linear and logarithmic sweeps are monotonously increasing, and a single sweep has no repetitions at any frequencies, while the MLS measurement is equivalent to repeating and averaging the impulses over the entire measurement time. If time variances occur within one measurement period, sweeps hardly reveal detrimental effects caused by time variances, while MLS show enormous errors [Svensson and Nielsen, 1999]. Secondly, the harmonic distortion artifacts that are excited by loudspeakers, can be easily discarded because these harmonic distortion occurs only at the very end of the measured impulse response, while the desired impulse response appears only at the beginning the impulse response. The distortion artifacts can then safely be windowed out.

In addition, if some narrow-band transient noise occurs occasionally, the spectrogram of the noise does not overlap with the sweeps and the noise can be windowed out as well. As shown in Fig. 2.1, the information of the impulse response lies only in the nearby time-frequency blocks that correspond to the delay curve, if the noise does not occur in those nearby time-frequency blocks of the sweep, the noise can be windowed out and may have no influence on the impulse response. For the MLS measurement, as long as the noise appears within the measurement period, the noise always has an influence on impulse response if no further signal processing is processed.

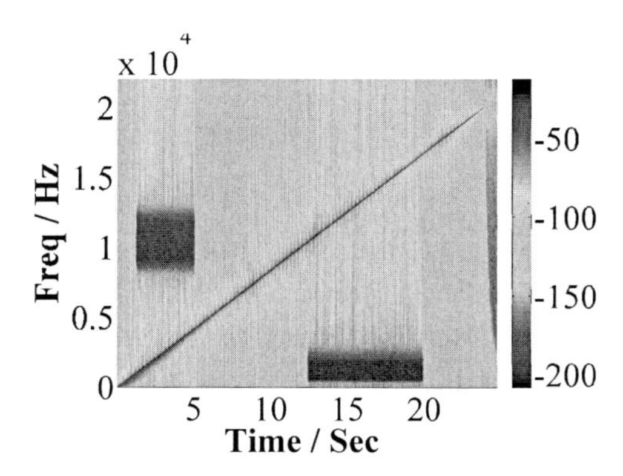

Figure 2.1.: The spectrogram of sweep and certain narrow-band transient noise

Furthermore, sweeps can also be designed flexibly to fit the specific noise spectrum. Compared with logarithmic sweeps and linear sweeps, the spectrum of a linear sweep is flat, and the spectrum of a logarithmic sweep deceases by 3 dB per octave. Logarithmic sweeps contain more energy at low frequencies than linear sweeps of the same length. Under general laboratory conditions or in room acoustics measurements, the random noise tends to involve more energy at low frequencies than at high frequencies.

In this case, the logarithmic sweep is a better choice because it carries more energy at low frequencies. However, if the random noise has a specific spectrum, it is also possible to adjust the swept-rate to an arbitrary spectrum distribution to compensate for the noise influence [Müller and Massarani, 2001]. For instance, if the noise spectrum is concentrated between 1,000 and 1,500 Hz and there is very low noise outside this range, the swept-rate, which is the growth rate of the group delay, could be chosen to be very slow in order to send more energy into this frequency range.

2.4. Multi-sensors based signal processing

As mentioned earlier, slowing down the swept rate—in other words, using long sweeps—can greatly improve SNR. Measurements in rooms such as factory halls require very long sweeps or a very long averaging time. Random noise caused by machines can be rejected as long as the swept rate is slow enough. However, if the measurement time is too long, the time-variance effect may appear. The multi-channels approach is described here assuming that the desired signal can be separated from the background noise during short-time measurements.

If a single noise source is present, one intuitive idea is to use two sensors. This is the standard Multiple Input Multiple Output (MIMO) problem, which is widely used in communication and speech enhancement techniques. Note that $s(t)$ and $n(t)$ are the excitation signal and the noise signal respectively. If the two sources are measured using two sensors, the recorded signal is the mixture of the convolution of the sources with the noises, as illustrated in Eq. 2.24

$$\begin{aligned} x_1(t) &= h_{11}(t) * s(t) + h_{12}(t) * n(t) \\ x_2(t) &= h_{21}(t) * s(t) + h_{22}(t) * n(t) \end{aligned} \tag{2.24}$$

where $s(t)$ is the excitation signal and $n(t)$ is the noise signal. $h_{ij}, (i = 1, 2 \quad j = 1, 2)$ are the corresponding impulse responses from the sources to the receivers. Written in the frequency domain, the relation between the sources and the receivers is Eq. 2.25

$$\begin{bmatrix} X_1(\omega) \\ X_2(\omega) \end{bmatrix} = \begin{bmatrix} H_{11}(\omega) & H_{12}(\omega) \\ H_{21}(\omega) & H_{22}(\omega) \end{bmatrix} \begin{bmatrix} S_1(\omega) \\ S_2(\omega) \end{bmatrix}$$
$$\mathbf{X} = \mathbf{HS} \tag{2.25}$$

Compared with the single-input single-output scenario, in Eq.2.16 and Eq.2.17, the problem becomes more complicated in the presence of noise. In the frequency domain, the transfer function H_{11} and H_{21} are measured. $S_1(\omega)$ is the excitation signal and $X_1(\omega)$ and $X_2(\omega)$ are the recorded signals from the microphones. $S_2(\omega)$ is the noise signal. If there is no noise, the H_{11}, H_{21} is obtained directly by dividing $X_1(\omega)$ and $X_2(\omega)$ by $S_1(\omega)$. When the noise $S_2(\omega)$ occurs, this becomes a ill-posed problem because there are only two equations but five unknown variables: $H_{11}(\omega)$, $H_{12}(\omega)$, $H_{21}(\omega)$, $H_{22}(\omega)$ and $S_2(\omega)$.

Generally, there are two ways to solve the above equations. One is spatial filtering and the other is the statistical method.

2.4.1. Spatial filtering

Narrowband spatial filtering

Spatial filtering is also called beamforming. The mathematical framework can be found as early as [Capon, 1969]. This technique has already been applied in various areas, such as speech enhancement and antenna engineering. The basic theory of beamforming is available in the books, [Dudgeon, 1993] and [Trees, 2002], and in the theory of directional microphones [Kuttruff, 2007]. The underlying idea is to pass the signal from a given direction and filter out the source from some another directions. Usually, the spatial filter treats the waves as far-field plane waves. Taking a two-microphone array as an example, the sound wave propagates through the array's aperture, and the signal recorded by the second sensor contains a delay with regard to the signal recorded by the first sensor. The array's output can be expressed as the sum of the two sensors with arbitrary weighing factors, as Eq. 2.26

$$y(t) = w_1 x(t) + w_2 x(t - kd\cos\theta) \tag{2.26}$$

where $x(t)$ and $x(t - kd\cos\theta)$ are the signals recorded by two sensors, and their difference is only a time delay $kd\cos\theta$. w_1 and w_2 are weighting factors. d is the distance between two microphones, k is the wave number, θ is the angle of the source. Considering a scenario, two sources appear simultaneously. S_1 and S_2 come from the angle θ_1 and θ_2 respectively, as shown in Fig.2.2. If the weighting factors are chosen by Eq. 2.27, the output will suppress the sound source from direction θ_2 and concentrate on the direction θ_1, as shown in Fig.2.2. The blue curve is the array directivity pattern. The source from θ_1 is reinforced and the source from θ_2 is suppressed.

$$\begin{aligned} w_1 &= \frac{-e^{-ikd\cos\theta_2}}{e^{-ikd\cos\theta_1} - e^{-ikd\cos\theta_2}} \\ w_2 &= \frac{1}{e^{-ikd\cos\theta_1} - e^{-ikd\cos\theta_2}} \end{aligned} \tag{2.27}$$

If a third noise source emerges from another direction, at least three microphones are needed.

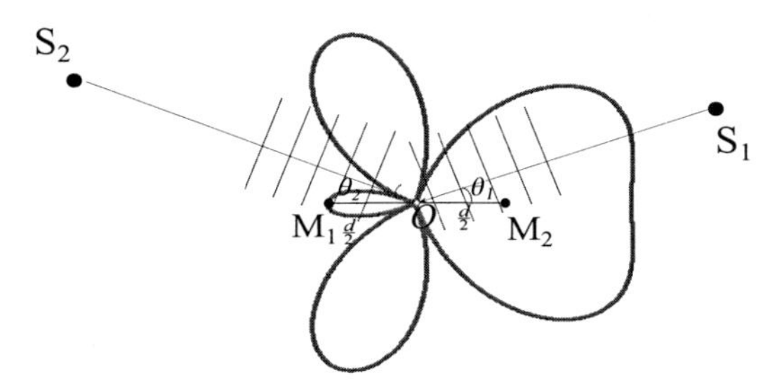

Figure 2.2.: Beamforming with two microphones

The generalized formulation can be written as follows. Considering that there are Q sources and P sensors. The sources vector is written as

$$\mathbf{S}(\omega) = \begin{bmatrix} s_1(\omega) & s_2(\omega) & \cdots & s_Q(\omega) \end{bmatrix}^T \tag{2.28}$$

The signal recorded by the p-th sensors is written as

$$\begin{aligned} x_p(\omega) = a_{p1}(\omega,\theta_1)s_1(\omega) + a_{p2}(\omega,\theta_2)s_2(\omega) + \cdots \\ + a_{pq}(\omega,\theta_Q)s_Q(\omega) + \cdots + a_{pQ}(\omega,\theta_Q)s_Q(\omega) \end{aligned} \tag{2.29}$$

where

$$a_{pq}(\omega,\theta_q) = e^{-ik\cdot d_{pq}\cdot\cos(\theta_q)} \tag{2.30}$$

and $p = 1, 2...P$, $q = 1, 2, ...Q$. d_{pq} is the distance from the q-th source to the p-th sensor.

Therefore, the overall signals recorded by the P sensors are written as

$$\begin{aligned} \begin{bmatrix} x_1(\omega) \\ x_2(\omega) \\ \vdots \\ x_P(\omega) \end{bmatrix} &= \begin{bmatrix} a_{11}(\omega,\theta_1) & a_{12}(\omega,\theta_2) & \cdots & a_{1Q}(\omega,\theta_Q) \\ a_{21}(\omega,\theta_1) & a_{22}(\omega,\theta_2) & \cdots & a_{2Q}(\omega,\theta_Q) \\ \vdots & \vdots & \ddots & \vdots \\ a_{P1}(\omega,\theta_1) & a_{P2}(\omega,\theta_2) & \cdots & a_{PQ}(\omega,\theta_Q) \end{bmatrix} \begin{bmatrix} s_1(\omega) \\ s_2(\omega) \\ \vdots \\ s_Q(\omega) \end{bmatrix} \\ \mathbf{X}(\omega) &= \mathbf{A}\mathbf{S}(\omega) \end{aligned} \tag{2.31}$$

If the number of sensors is equal to the number of sources ($P = Q$), the sources can be recovered by multiplying the inverse of the mixing matrix $\mathbf{A}$.

$$\mathbf{S}(\omega) = \mathbf{A}^{-1}\mathbf{X}(\omega) \tag{2.32}$$

If the number of sensors is larger than the sources ($P > Q$)

$$\mathbf{S}(\omega) = \mathbf{A}^{+}\mathbf{X}(\omega) \tag{2.33}$$

where $\mathbf{A}^{+}$ denotes the Moore–Penrose pseudoinverse $\mathbf{A}^{+} = (\mathbf{A}^{H}\mathbf{A})^{-1}\mathbf{A}^{H}$. Once the directions of the noise sources are known, the original signal can be recovered by computing the inverse matrix of $\mathbf{A}$.

If the sources to be treated are close to the sensors, the near-field spherical harmonics must be considered. [Kennedy, 1998] [Kennedy, 1999] [Abhayapala, 2000] [Ryan, 2000]

Broadband optimization

The equations of spatial filtering described above express only a single frequency. It cannot be applied for broadband directly because the matrix A in Eq.2.31 might be ill-conditioned at certain frequencies.

In order to pick up broadband signals from different directions, numerous methods have been investigated over a long period and are described using different terminologies. Unequally spaced arrays derived by the asymptotic theory are developed by [Ishimaru, 1962] [Ishimaru, 1965], and the resulting arrays have exponential spacing. [Chou, 1995a] [Chou, 1995b] unitizes the nested array to keep the null directions frequency-constant over audio frequencies (500 $\sim$ 7200 Hz). A generalized mathematical framework to build a frequency-invariant beam pattern for far-field sources is introduced by [Darren B. Ward and Williamson, 1995] and is based on the continuous array aperture. It represents the outputs of the array and the array aperture by integral equation. The array aperture for a desired beam pattern is the solution of this integral equation. The broadband near-field beam pattern design is addressed by [Kennedy, 1998], which uses the spherical harmonic solution to transform the near-field beam pattern to an equivalent far-field beam pattern. When the number of sensors is restricted and the array aperture is arbitrarily configured, a condition number can be used as a criterion for the frequency bands when computing the array response [Parra, 2006].

The condition number is defined to quantify the numerical error of computing the inverse matrix [E. Ward Cheney, 2007]. Considering the solution of a linear system,

$$\mathbf{A}\mathbf{x} = \mathbf{b} \tag{2.34}$$

The definition of condition number is

$$\kappa(A) = \| \mathbf{A} \| \| \mathbf{A}^{-1} \| \tag{2.35}$$

If the input $\mathbf{b}$ contains a little noise $\delta\mathbf{b}$, the error of the output is

$$\frac{\| \delta\mathbf{x} \|}{\| \mathbf{x} \|} \leq \kappa(A) \frac{\| \delta\mathbf{b} \|}{\| \mathbf{b} \|} \tag{2.36}$$

where the symbol $\leq$ denotes the error bound of the estimation. When the condition number is large, the solution of the system may have large errors. When the condition number is close to one, the matrix is well conditioned. The identity matrix's condition number is one.

Since the elements of the mixing matrix A in Eq.2.31 are related to array microphone-to-microphone distance and the wave-incidence angle, optimizing the array to cover broadband frequency range is equivalent to choosing a suitable array aperture and making A well conditioned for the broadband frequency range in specific directions.

Spatial filtering could be applied in two scenarios in principle.

1. A few sources are located in different directions and each source has only direct sound but no reflections. Spatial filtering can be implemented to separate the sources from different directions.

2. The sources have a limited number of reflections. Spatial filtering separates the waves from different directions. These sound waves can be the reflected sound of the identical source as well as the direct sound or reflected sound of a different source.

The limitation of the spatial filter is that the number of sensors must be no less than the number of directions to be separated because there is no solution for A in Eq. 2.31 if the number of directions are larger than the number of sensors. If the scenario is a reverberant system, spatial filtering fails. However, the statistics-based methods could be implemented to separate different sources.

2.4.2. Statistics-based methods

General solution for linear equations with statistics based methods

The generalized mathematical framework of the statistics-based methods is the mixture as given in Eq.2.37.

$$\begin{bmatrix} x_1 \\ x_2 \end{bmatrix} = \begin{bmatrix} a_{11} & a_{12} \\ a_{21} & a_{22} \end{bmatrix} \begin{bmatrix} s_1 \\ s_2 \end{bmatrix}$$
$$\mathbf{X} = \mathbf{AS} \tag{2.37}$$

The goal is to extract s_1 and s_2 from the observations x_1 and x_2. This seems to be an ill-posed problem and has no solution because the mixing matrix $\mathbf{A}$ is unknown and s_1 and s_2 are also unknown. The number of equations is smaller than the number of parameters. As early as in 1988, however, Lacoume [Lacoume, 1988] has shown that by introducing the fourth-order cumulants as the supplementary equations, the problem Eq.2.37 becomes solvable in approximation.

The only assumption is that the two sources are independent and that, at most, only one source is Gaussian noise. Later on, the terminology Independent Component Analysis (ICA) is used by [Comon, 1994]. Thereafter, the ICA has become a very popular topic in signal processing over the past two decades. Thousands of papers are published based on the theory and applications of ICA.

The principle of ICA can be briefly described as follows. Concerning the unmixing matrix $\mathbf{W}$, a contrast function $f(\mathbf{WX})$ must be found by minimizing or maximizing the contrast function $f(\mathbf{WX})$. This results in the extraction of the sources s_1, s_2. Actually, these contrast functions are the supplementary functions with extraneous constraints based on the statistical characters of the signals, which helps to solve the ill-posed equation Eq. 2.37.

For most of the ICA algorithms, the contrast function does not exactly solve all the elements of matrix $\mathbf{A}$ in Eq. 2.37. The scale and permutation ambiguity still cause uncertainty. As shown in Eq. 2.38, an unmixing matrix $\mathbf{W}$ is found, the overall response of the unmixing process is $\mathbf{WA}$, and if an overall response is a diagonal or anti-diagonal matrix (Eq. 2.39), the source is extracted. Even if the magnitude of α_1, α_2 cannot be solved, and the permutation is unknown, it is still acceptable for many applications.

$$\begin{aligned} \mathbf{Y} &= \mathbf{WX} = \mathbf{WAS} \\ &= \mathbf{\Lambda S} \end{aligned} \tag{2.38}$$

where

$$\mathbf{\Lambda} = \begin{bmatrix} \alpha_1 & 0 \\ 0 & \alpha_2 \end{bmatrix} \text{ or } \begin{bmatrix} 0 & \alpha_1 \\ \alpha_2 & 0 \end{bmatrix} \tag{2.39}$$

For example, in speech enhancement, the absolute energy level in a person's voice is not as important. Some other a priori information may be used to determine which voice belongs to whom.

Generally, statistical characteristics that are chosen to build the contrast function are non-Gaussianity, maximum likelihood or mutual information, etc. They are described extensively in the book [Hyvarinen, 2001]. The criterion of non-Gaussianity is briefly explained here through an example. The principle of the non-Gaussianity estimator is the central limit theorem: if s_1 and s_2 are non-Gaussian independent and identically distributed random variables, the sum of s_1 and s_2 is closer to a Gaussian distribution than are $s_1$and s_2 individually. In order to extract the original source s_1 and s_2, a linear combination is operated on $\mathbf{x}$, $y = \sum_{i=1}^{2} b_i x_i$,

$$y = \mathbf{b}^T\mathbf{x} = \mathbf{b}^T\mathbf{A}\mathbf{s} = \sum_{i=1}^{2} c_i s_i \tag{2.40}$$

According to the central limit theorem, y should be closer to the Gaussian distribution than s_1 and s_2 individually with the only two exceptions being $c_1 = 0, c_2 \neq 0$ or $c_1 \neq 0, c_2 = 0$.

In order to describe the terms 'closer to Gaussian distribution' or 'non-Gaussianity' correctly, [Comon, 1994] uses negentropy, which is defined as,

$$J(y) = H(y_{Gauss}) - H(y) \tag{2.41}$$

where $H(y)$ is the entropy function of a random vector y with probability density $p_y(y)$, and y_{Gauss} is the Gaussian variable that has identical covariance to y.

$$H(y) = -\int p_y(y) \log p_y(y) \mathbf{d}y \tag{2.42}$$

$p_y(y)$ is the probability distribution of y.

Because the Gaussian variables have the largest entropy among all the random variables, the central limit theorem can be explained by

$$J(y) \leq J(s_1) \text{ or } J(s_2) \tag{2.43}$$

The equality holds only if $y = s_1$ or $y = s_2$. Therefore, the original independent signal can be recovered by maximizing $J(y)$. $J(y)$ can be approximated as

$$J(y) \propto \Big[E\left\{G(y)\right\} - E\left\{G(v)\right\}\Big] \tag{2.44}$$

where

$$G(y) = \frac{1}{a_1} \log \cosh a_1 y, \text{ and } 1 \leq a_1 \leq 2 \tag{2.45}$$

If the sources are two non-Gaussian independent sources, the contrast function has two maxima, with each maximum belonging to one individual non-Gaussian independent source. If the sources are two Gaussian sources, since the sum of the two Gaussian variables is still a Gaussian variable, this non-Gaussianity contrast function cannot distinguish two Gaussian sources.

For inequality of non-Gaussianity Eq. 2.43, the number of the sources is not limited to two. If more than two sources exist, this inequality still holds. In this case, the number of sensors must be no less than the number of sources. The ICA is processed using the following three steps.

1. Whitening

If the sources are independent, they must be uncorrelated. When the sources are extracted from the mixture, the extracted signals must be uncorrelated with each other. The first step of ICA is decorrelating the mixed signals. The mixed signals are decorrelated by Singular Value Decomposition (SVD) [Strang, 1988].

$$\mathbf{U}\mathbf{\Sigma}\mathbf{V}^T = \mathbf{x}\mathbf{x}^T \tag{2.46}$$

$$\mathbf{z} = \mathbf{\Sigma}^{-1/2}\mathbf{U}^T\mathbf{x} \tag{2.47}$$

where $\mathbf{x}$ is a zero-mean random vector and $\mathbf{z}$ is a unit-variance uncorrelated random vectors. The purpose of this step is to constrain the separated signals to be uncorrelated and reduce the computing power of the estimation.

2. Maximizing the Negentropy function

If $\mathbf{z}$ is multiplied by any rotation matrix, as Eq. 2.48, $\mathbf{y}$ remains uncorrelated, but the negentropy of the individual vector will be changed.

$$\begin{aligned}\mathbf{Y} = \begin{bmatrix} y_1 \\ y_2 \end{bmatrix} &= \begin{bmatrix} \cos\theta & -\sin\theta \\ \sin\theta & \cos\theta \end{bmatrix} \mathbf{z} \\ &= \mathbf{G}(\theta)\mathbf{z} \end{aligned} \tag{2.48}$$

Therefore, by maximizing the negentropy (Eq. 2.44) of y_1 and y_2, the independent sources are extracted. [1]

At this moment the permutation of the source and the absolute magnitude of the sources are still unknown. $\mathbf{Y} = \begin{bmatrix} \alpha_1 & 0 \\ 0 & \alpha_2 \end{bmatrix} \mathbf{s}$ or $\mathbf{Y} = \begin{bmatrix} 0 & \alpha_1 \\ \alpha_2 & 0 \end{bmatrix} \mathbf{s}$

If the absolute value is required, a priori information must first be used to determine the permutation of the sources, and ascertain which signal belongs to which source. The magnitude of the signals can then be obtained by least mean square estimation.

3. Magnitude Estimation

Assuming that the two sources are recovered, that the permutation of the separated source has been known by some other prior knowledge, and only the magnitude is unknown, the separated signals are $\mathbf{Y} = \begin{bmatrix} \alpha_1 & 0 \\ 0 & \alpha_2 \end{bmatrix} \begin{bmatrix} s_1 \\ s_2 \end{bmatrix}$

The relation between the separated signal $\mathbf{Y}$ and the measured signal $\mathbf{X}$ is

$$\begin{aligned}\mathbf{X} &= \mathbf{BY} \\ &= \mathbf{B\Lambda S} \\ &= \mathbf{AS} \end{aligned} \tag{2.49}$$

$\mathbf{B}$ can be calculated by the Moore–Penrose inverse:

[1] In order to find the maximum of the contrast function, the general method used is gradient descent. The details of this method can be found in any book on mathematical optimization. For example, [Snyman, 2005]

$$\mathbf{B} = \mathbf{X}\mathbf{Y}^{\mathrm{H}} \left(\mathbf{Y}\mathbf{Y}^{\mathrm{H}}\right)^{+} \tag{2.50}$$

where $\mathbf{B}$ is a 2×2 matrix. Then $\mathbf{C} = \mathbf{B}\mathbf{Y}$, and every element of the $\mathbf{C}$ includes the absolute value of the sources arriving at the sensors.

The essential step for ICA is the second step—maximizing the negentropy function. If the sources are correlated, they cannot be separated by maximizing the negentropy function and the third step, magnitude estimation, cannot be guaranteed as well.

The above description of ICA is defined for the real-valued signals. If the signal is represented in the frequency domain, the variables are complex-valued numbers. Since every complex number has two dimensions—the real part and the imagery part—the statistical characteristics between two complex-valued numbers differ slightly from those of the real number statistics, and the formulation of ICA has to be changed. The mathematical framework of circular and non-circular sources is explained in the book [Schreier, 2010].

There are two different definitions on the covariance of two signals. These are the covariance matrix and pseudo covariance matrix.[Neeser, 1993] [Eriksson, 2004].

The definition of covariance matrix is

$$R_{ss} = E\left\{\mathbf{s}\mathbf{s}^{H}\right\} \tag{2.51}$$

The definition of pseudo covariance matrix is

$$\widetilde{R}_{ss} = E\left\{\mathbf{s}\mathbf{s}^{T}\right\} \tag{2.52}$$

The complex-valued ICA requires that the two sources are uncorrelated with each other, which means $E\left\{\mathbf{s}\mathbf{s}^{H}\right\} = \mathbf{I}$. In the meanwhile, the pseudo-covariance matrix must be zeros $E\left\{\mathbf{s}\mathbf{s}^{T}\right\} = \mathbf{0}$ [Bingham, 2000]. The complex-valued ICA is then performed by the following three steps.

1. Whitening

In case the sources are second-order circular $E\left\{\mathbf{s}\mathbf{s}^T\right\} = \mathbf{0}$, the normal SVD is performed on the covariance matrix of $\mathbf{X}$,

$$\mathbf{U\Sigma V}^H = \mathbf{XX}^H \tag{2.53}$$

where $\mathbf{\Sigma} = \begin{bmatrix} \sigma_1 & 0 \\ 0 & \sigma_2 \end{bmatrix}$ is a diagonal matrix. Then two orthogonal whitened vectors $\mathsf{y}_1, \mathsf{y}_2$ are obtained by Eq. 2.54.

$$\begin{bmatrix} \mathsf{y}_1 \\ \mathsf{y}_2 \end{bmatrix} = \mathbf{Y} = \begin{bmatrix} 1/\sigma_1 & 0 \\ 0 & 1/\sigma_2 \end{bmatrix} \mathbf{U}^H \mathbf{X} \tag{2.54}$$

[Douglas, 2007] use Strong Uncorrelating Transform (SUT) instead of SVD as the whitening process in order to achieve the necessary robustness.

2. Maximizing the Non-Gaussianity

Since all the variables used here are complex numbers, the rotation matrix is written in the complex domain as well. This rotation matrix is taken as follows, which is also used by [Belouchrani, 1997].

$$\mathbf{G}(\theta, \phi) = \begin{bmatrix} \cos\theta & e^{i\phi}\sin\theta \\ -e^{-i\phi}\sin\theta & \cos\theta \end{bmatrix} \tag{2.55}$$

This rotation matrix helps to find the solution of the Eq. 4.7.

$$\begin{aligned} \begin{bmatrix} \mathsf{z}_1 \\ \mathsf{z}_2 \end{bmatrix} = \mathbf{Z} &= \mathbf{G}(\theta, \phi)\mathbf{Y} \\ &= \mathbf{G}(\theta, \phi)\mathbf{U}^H\mathbf{X} \\ &= \mathbf{G}(\theta, \phi)\mathbf{U}^H\mathbf{HS} \\ &= \mathbf{\Lambda S} \end{aligned} \tag{2.56}$$

If a suitable parameter pair(θ,ϕ) is chosen to create the overall response $\mathbf{\Lambda}$ matrix diagonal, which is

$$\begin{bmatrix} \mathsf{z}_1 \\ \mathsf{z}_2 \end{bmatrix} = \begin{bmatrix} \alpha_1 & 0 \\ 0 & \alpha_2 \end{bmatrix} \begin{bmatrix} \mathsf{s}_1 \\ \mathsf{s}_2 \end{bmatrix} \tag{2.57}$$

then the sources are separated. The contrast function to determine the suitable (θ,ϕ) is the approximation of the non-Gaussianity.

$$J(|y|) \propto \left[E\left\{G(|y|)\right\} - E\left\{G(|v|)\right\} \right] \tag{2.58}$$

Compared with the real-valued non-Gaussianity function Eq. 2.44, the absolute value must be taken in Eq. 2.58. Eq. 2.45 can be used as an approximation of G(y). [Bingham, 2000] suggests that $G(y) = \sqrt{(a_1 + y)}$, $G(y) = \sqrt{0.1 + y}$, $G(y) = \log(0.1 + y)$ or $G(y) = \frac{1}{2}y^2$ is robust against outliers. The maximum likelihood and mutual information minimization can also be used to separate the complex-valued sources [Adali, 2008].

Another robust contrast function is the joint diagonalization [Cardoso, 1993] [Belouchrani, 1997], which uses the cross-correlation function as the contrast function. The two signals can be either real-valued or complex-valued. Assuming that the two sources are incoherent, the cross-correlation function between the two signals is zero (Eq. 2.60).

$$\mathbf{s(t)} = \begin{bmatrix} s_1(t) \\ s_2(t) \end{bmatrix} \tag{2.59}$$

$$E\left[\mathbf{s(t)}\mathbf{s(t}+\tau)^H \right] = \begin{bmatrix} \rho_1(\tau) & 0 \\ 0 & \rho_2(\tau) \end{bmatrix} \tag{2.60}$$

where $\rho_i(\tau) = E\left[s_i(t) \cdot s_i(t+\tau)^*\right]$

Regarding a signal mixture $\mathbf{X}(\mathbf{t}) = \mathbf{As}(\mathbf{t})$, the cross-covariance matrix (Eq. 2.62) can be used as the contrast function to separate the sources.

$$\mathbf{Z}(\mathbf{t}) = \mathbf{WX}(\mathbf{t}) \tag{2.61}$$

$$R(\mathbf{W}, \tau) = E\left[\mathbf{Z}(\mathbf{t})\mathbf{Z}(\mathbf{t}+\tau)^{H}\right] \tag{2.62}$$

Thus, the contrast function is

$$\text{Off}\left[R(\mathbf{W}, \tau)\right] = \sum_{\tau}\left[\sum_{i,j.i\neq j}\left(\left|R_{i,j}(\mathbf{W}, \tau)\right|^{2}\right)\right] \tag{2.63}$$

When a suitable $\mathbf{W}$ is chosen to diagonalize $R(\mathbf{W}, \tau)$ for all the τs and to minimize Eq. 2.63, the two sources are separated.

The linear mixture model described above requires that the sources are simultaneously mixed. In case the signals of sources arrive at the sensors with several different delays—i.e. in a convolutive mixture—the above linear mixture model fails to separate the sources. The contrast function for this convolutive mixture must be modified. Theoretically, a convolutive mixture can be treated in either the frequency domain or the time domain. In the frequency domain, the time-domain convolutive mixture becomes an instantaneous mixture, and the complex-valued ICA or joint diagonalization can be applied.

In the time domain, this is a MIMO inverse FIR filter problem. The mathematical framework for the time-domain convolutive mixture was described by [Comon, 1996] [Moreau, 1997] [Comon, 2003]. A generalized mathematical framework is expressed by [Buchner, 2005]: the goal is the replacement of the contrast function of diagonal matrix by block diagonal matrix.

The measured signals of the convolutive mixture are

$$
\begin{aligned}
x_1(t) &= h_{11}(t) * s_1(t) + h_{12}(t) * s_2(t) \\
x_2(t) &= h_{21}(t) * s_1(t) + h_{22}(t) * s_2(t)
\end{aligned}
\tag{2.64}
$$

[Buchner, 2005] formulates the convolutive mixture as the block-wise matrix of FIR filter Eq. 2.65,

$$
h_{pq}(t) * s_q(t) = \begin{bmatrix}
h_{pq,0} & 0 & 0 & \cdots & 0 \\
h_{pq,1} & h_{pq,0} & 0 & \cdots & 0 \\
\vdots & h_{pq,1} & h_{pq,0} & \cdots & 0 \\
h_{pq,M-1} & \vdots & h_{pq,1} & \cdots & 0 \\
0 & h_{pq,M-1} & \vdots & \ddots & 0 \\
\vdots & 0 & h_{pq,M-1} & \ddots & h_{pq,0} \\
0 & \vdots & 0 & \ddots & h_{pq,1} \\
0 & 0 & \vdots & \ddots & \vdots \\
0 & 0 & 0 & \cdots & h_{pq,M-1}
\end{bmatrix}
\begin{bmatrix} s_q(0) \\ s_q(1) \\ \vdots \\ \vdots \\ s_q(N) \end{bmatrix} = \mathbf{H}_{pq}\mathbf{s}_q
\tag{2.65}
$$

where M is filter order. The overall mixing mixture is

$$
\mathbf{X} = \widehat{\mathbf{H}}\mathbf{S} = \begin{bmatrix} \mathbf{H}_{11} & \mathbf{H}_{12} \\ \mathbf{H}_{21} & \mathbf{H}_{22} \end{bmatrix} \begin{bmatrix} \mathbf{s}_1 \\ \mathbf{s}_2 \end{bmatrix}
\tag{2.66}
$$

The ideal separation is to find out an unmixing block-wise matrix $\widehat{\mathbf{W}}$

$$
\widehat{\mathbf{W}} = \begin{bmatrix} \mathbf{W}_{11} & \mathbf{W}_{12} \\ \mathbf{W}_{21} & \mathbf{W}_{22} \end{bmatrix}
\tag{2.67}
$$

which makes $\widehat{\mathbf{C}} = \widehat{\mathbf{W}}\widehat{\mathbf{H}}$ a perfect diagonal matrix. But this is a challenging task because the perfect inverse of the block-wise matrix $\widehat{\mathbf{H}}$ does not always exist. However, another looser solution is to make the recovered matrix $\widehat{C}$ block diagonal, which is

$$\widehat{\mathbf{C}}_{\text{Block diag}} = \widehat{\mathbf{W}}\widehat{\mathbf{H}} = \begin{bmatrix} \mathbf{C}_{11} & \mathbf{0} \\ \mathbf{0} & \mathbf{C}_{22} \end{bmatrix} \tag{2.68}$$

Then the output signal is

$$\begin{bmatrix} \mathbf{y}_1 \\ \mathbf{y}_2 \end{bmatrix} = \begin{bmatrix} \mathbf{C}_{11} & \mathbf{0} \\ \mathbf{0} & \mathbf{C}_{22} \end{bmatrix} \begin{bmatrix} \mathbf{s}_1 \\ \mathbf{s}_2 \end{bmatrix} = \begin{bmatrix} \mathbf{C}_{11}\mathbf{s}_1 \\ \mathbf{C}_{22}\mathbf{s}_2 \end{bmatrix} \tag{2.69}$$

The output signal $\mathbf{y}_1, \mathbf{y}_2$ is the filtered version of $\mathbf{s}_1, \mathbf{s}_2$ by an arbitrary filter $\mathbf{C}_{11}, \mathbf{C}_{22}$. This is still a good solution if the application is speech enhancement[Kellermann, 2004] [Kellermann, 2010] because the noise is separated from the source in principle. If the application is impulse response measurement, however, the results, including the distortion artifacts, is unacceptable.

2.4.3. Comparison with averaging

The source separation method must be compared with the conventional synchronous average to evaluate its performance. There is a 3 dB rule for averaging the signal in a time-invariant system. When the signal is averaged twice, the SNR is improved by 3 dB. This rule can be derived as follows:

When the measurements are contaminated with noise, the measured signal is,

$$X_i(\omega) = H(\omega)S_i(\omega) + N_i(\omega) \tag{2.70}$$

To reduce the measurement error, the measurement is repeated m times, as

$$\bar{X}(\omega) = \frac{1}{m}\sum_{i}^{m} X_i(\omega) \tag{2.71}$$

The SNR improvement of the average method is actually the principle of standard deviation of the sample mean in statistics, which is

$$\sigma = \sqrt{\frac{1}{m}\sum_{i=1}^{m}\left(X_i(\omega) - \bar{X}(\omega)\right)^2} \tag{2.72}$$

If the signal is averaged m times, the standard deviation of the estimation error is reduced by $\frac{1}{\sqrt{m}}$, which is equivalent to SNR being increased by $\sqrt{m}$ or $10\log_{10} m$ dB.

Averaging is actually equivalent to the popular terminology Least-Square Estimation (LSE) (Eq. 2.74), and to the theory of LSE, which is described in detail in the books on statistics [Barkat, 2005] [Maurice George Kendall, 1961] [Franklin A. Graybill, 1994][Weisberg, 2005]. Considering that there are p sources, q sensors and m measurements $(m > p)$, the measured signals are

$$\mathbf{X} = \mathbf{HS} + \mathbf{N} \tag{2.73}$$

where $\mathbf{X}$ is the measured matrix of $q \times m$ matrix and $\mathbf{S}$ denotes the known sources, which is a $p \times m$ matrix. $\mathbf{N}$ is the noise. The LSE of $\mathbf{H}$ is

$$\hat{\mathbf{H}} = \mathbf{XS}^H\left(\mathbf{SS}^H\right)^{-1} \tag{2.74}$$

If there is only one source $p = 1$ and only one sensor $q = 1$, and the observation of the sources $\mathbf{S}$ is just the repetition of a constant signal. The 2.74 is simplified to the equation of average Eq. 2.71.

Furthermore, another general rule for the parameter estimation must be considered. Once the samples $s_1, s_2, \cdots, s_m$ and $x_1, x_2, \cdots, x_m$ are measured, the system is modeled as

$$\mathbf{X} = f(s_1, s_2, \cdots, s_m, \beta_1, \beta_2, \cdots, \beta_l) \tag{2.75}$$

where l is the number of unknown parameters and m is the number of measured samples. The model $f(s_1, s_2, \cdots, s_m, \beta_1, \beta_2, \cdots, \beta_l)$ can be either linear or nonlinear. The estimation error of standard deviation is [Franklin A. Graybill, 1994].

$$\sigma \propto \sqrt{\frac{1}{m - l - 1}} \tag{2.76}$$

The more samples used to perform the estimation, the fewer the errors. On the other hand, the greater the number of estimated parameters, the larger the errors in the estimation.

3

Separation of direct sound and reflected sound: Reflection coefficient in situ measurement

The spatial filter can separate the signal from different directions assuming that the wave is a plane wave [Darren B. Ward and Williamson, 1995] or known spherical harmonics [Kennedy, 1998]. In terms of reflection coefficient measurement outdoors, however, the microphones are very close to the loudspeakers and the surface under test. When the wave front is neither a plane wave nor a known spherical harmonics, the magnitude of the signal impinging on different microphones should be estimated. In this chapter, the general solution and broadband optimization for the separation of sounds from two different directions are described first. Then, the possible application of reflection coefficient measurement is investigated.

3.1. General solution to separate direct sound and reflected sound

The signals recorded by the two sensors should be written as Eq.3.1.

$$\begin{aligned}
\text{Mic 1}: \quad & x_1(t) = s_1(t) && + \quad s_2(t) \\
\text{Mic 2}: \quad & x_2(t) = h_{21}s_1(t-\tau_1) && + \quad h_{22}s_2(t-\tau_2)
\end{aligned} \tag{3.1}$$

where $s_1(t)$ and $s_2(t)$ are signals from two different directions, and τ_1 and τ_2 are assumed to be known. If the h_{21} and h_{22} are assumed to be approximately equal to one, the conventional far-field beamforming can be designed to be null in the unwanted direction and maximum in the wanted direction. But the magnitudes h_{21} and h_{22} are usually unequal and unknown in practice. Eq. 3.1 is actually an undetermined problem. There are four unknown parameters $h_{21}, h_{22}, s_1(t)$ and $s_2(t)$, and only two equations.

A practical solution of h_{21} and h_{22} arbitrarily uses more sensors to add supplementary equations and to make the equations determined. Considering that four microphones are used, the measured signal is expressed as Eq. 3.2.

$$\begin{array}{llccc} \text{Mic } 1: & x_1(t) = & s_1(t) & + & s_2(t) \\ \text{Mic } 2: & x_2(t) = & h_{21}\cdot s_1(t-\tau_{21}) & + & h_{22}\cdot s_2(t-\tau_{22}) \\ \text{Mic } 3: & x_3(t) = & h_{31}\cdot s_1(t-\tau_{31}) & + & h_{32}\cdot s_2(t-\tau_{32}) \\ \text{Mic } 4: & x_4(t) = & h_{41}\cdot s_1(t-\tau_{41}) & + & h_{42}\cdot s_2(t-\tau_{42}) \end{array} \tag{3.2}$$

Eq. 3.2 is transformed into the frequency domain for convenience.

$$\begin{bmatrix} X_1(\omega) \\ X_2(\omega) \\ X_3(\omega) \\ X_4(\omega) \end{bmatrix} = \begin{bmatrix} 1 & 1 \\ h_{21}e^{-i\omega\tau_{21}} & h_{22}e^{-i\omega\tau_{22}} \\ h_{31}e^{-i\omega\tau_{31}} & h_{32}e^{-i\omega\tau_{32}} \\ h_{41}e^{-i\omega\tau_{41}} & h_{42}e^{-i\omega\tau_{42}} \end{bmatrix} \begin{bmatrix} S_1(\omega) \\ S_2(\omega) \end{bmatrix}$$
$$\mathbf{X} = \mathbf{HS} \tag{3.3}$$

Although Eq. 3.3 is still an undetermined problem, the two additional microphones provide a way to construct a contrast function to estimate the parameters h_{ij}.

Note that

$$\mathsf{X}_{12} = \begin{bmatrix} X_1(\omega) \\ X_2(\omega) \end{bmatrix} = \begin{bmatrix} 1 & 1 \\ h_{21}e^{-i\omega\tau_{21}} & h_{22}e^{-i\omega\tau_{22}} \end{bmatrix} \begin{bmatrix} S_1(\omega) \\ S_2(\omega) \end{bmatrix}$$
$$= \mathsf{H}_{12}\mathbf{S} \tag{3.4}$$

$$\mathsf{X}_{34} = \begin{bmatrix} X_3(\omega) \\ X_4(\omega) \end{bmatrix} = \begin{bmatrix} h_{31}e^{-i\omega\tau_{31}} & h_{32}e^{-i\omega\tau_{32}} \\ h_{41}e^{-i\omega\tau_{41}} & h_{42}e^{-i\omega\tau_{42}} \end{bmatrix} \begin{bmatrix} S_1(\omega) \\ S_2(\omega) \end{bmatrix} = \mathsf{H}_{34}\mathbf{S} \tag{3.5}$$

If two unmixing matrices W_{12} and W_{34} can be found to separate the sources as Eq. 3.6 and Eq. 3.7,

$$\mathsf{W}_{12}\mathsf{H}_{12}\mathbf{S} = \begin{bmatrix} \mathsf{y}_1 \\ \mathsf{y}_2 \end{bmatrix} = \begin{bmatrix} c_1 & 0 \\ 0 & c_2 \end{bmatrix} \begin{bmatrix} S_1(\omega) \\ S_2(\omega) \end{bmatrix} \text{ or } \begin{bmatrix} 0 & c_1 \\ c_2 & 0 \end{bmatrix} \begin{bmatrix} S_1(\omega) \\ S_2(\omega) \end{bmatrix} \tag{3.6}$$

$$\mathsf{W}_{34}\mathsf{H}_{34}\mathbf{S} = \begin{bmatrix} \mathsf{y}_3 \\ \mathsf{y}_4 \end{bmatrix} = \begin{bmatrix} c_3 & 0 \\ 0 & c_4 \end{bmatrix} \begin{bmatrix} S_1(\omega) \\ S_2(\omega) \end{bmatrix} \text{ or } \begin{bmatrix} 0 & c_3 \\ c_4 & 0 \end{bmatrix} \begin{bmatrix} S_1(\omega) \\ S_2(\omega) \end{bmatrix} \tag{3.7}$$

then y_1 and y_3 should be perfectly correlated in the time domain whose correlation is one; otherwise, the correlation will be smaller than one, as per Eq. 3.8. Therefore, the source can be separated by maximizing the contrast function Eq. 3.8.

$$\text{Corr}\left(\mathsf{y}_1, \mathsf{y}_3\right) \leq 1 \tag{3.8}$$

On formulating the solution in this way, two uncertainties still remain.

1. Permutation ambiguity: Maximizing the contrast function Eq. 3.8 gives no information on which separated signal belongs to which source because the overall response $\begin{bmatrix} c_1 & 0 \\ 0 & c_2 \end{bmatrix}$ or $\begin{bmatrix} 0 & c_1 \\ c_2 & 0 \end{bmatrix}$ cannot be determined. This permutation ambiguity can be determined by the a priori information on the direction of arrival (Direction of Arrival (DOA))[Dudgeon, 1993].

2. Ill-condition problem: The mixing matrices H_{12} and H_{34} in Eq. 3.6 and Eq. 3.7 might be ill-conditioned at certain frequencies. The array aperture should be optimized to cover broadband signals.

3.2. Practical array optimization

Implementing the procedure from Eq. 3.3 to Eq. 3.8 enables the separation of sounds from two different directions. However, $\mathbf{H}$ or H_{12} and H_{34} might be ill-conditioned in practice at particular frequencies and cannot be accurately inverted. [Parra, 2006] uses the matrix regularization approach to improve the stability of the matrix inverse for microphone arrays, and uses the condition number as a criterion to exclude the frequencies of an ill-conditioned mixing matrix when computing the overall array responses. The condition number is used here as a criterion to optimize the array aperture and to make the mixing matrix well-conditioned for all frequencies. It is difficult to optimize the array by treating the H_{12} and H_{34} separately. An alternative process to optimize the condition number is presented below.

The four channels' signal in Eq. 3.3 is aligned by arbitrary delay as

$$\begin{aligned} \text{Ch } 1 &= X_1(\omega) \\ \text{Ch } 2 &= X_2(\omega) \cdot e^{i\omega(\tau_{21})} \\ \text{Ch } 3 &= X_3(\omega) \cdot e^{i\omega(\tau_{21}-\tau_{22}+\tau_{32})} \\ \text{Ch } 4 &= X_4(\omega) \cdot e^{i\omega(\tau_{21}-\tau_{22}+\tau_{32}-\tau_{31}+\tau_{41})} \end{aligned} \tag{3.9}$$

The aligned signal is

$$\begin{bmatrix} \text{Ch } 1 \\ \text{Ch } 2 \\ \text{Ch } 3 \\ \text{Ch } 4 \end{bmatrix} = \begin{bmatrix} 1 & 1 \\ h_{21} & h_{22}e^{i\omega(\tau_{21}-\tau_{22})} \\ h_{31}e^{i\omega(\tau_{21}-\tau_{22}+\tau_{32}-\tau_{31})} & h_{32}e^{i\omega(\tau_{21}-\tau_{22})} \\ h_{41}e^{i\omega(\tau_{21}-\tau_{22}+\tau_{32}-\tau_{31})} & h_{42}e^{i\omega(\tau_{21}-\tau_{22}+\tau_{32}-\tau_{31}+\tau_{41}-\tau_{42})} \end{bmatrix} \begin{bmatrix} S_1(\omega) \\ S_2(\omega) \end{bmatrix} \tag{3.10}$$

Shifting the delay of Eq.3.10 to Eq. 3.11

$$\begin{bmatrix} \text{Ch 1} \\ \text{Ch 2} \\ \text{Ch 3} \\ \text{Ch 4} \end{bmatrix} =$$

$$\begin{bmatrix} 1 & 0 & 1 & 0 \\ h_{21} & 0 & 0 & h_{22} \\ 0 & h_{31} & 0 & h_{32} \\ 0 & h_{41} & h_{42}e^{i\omega(\tau_{21}-\tau_{22}+\tau_{32}-\tau_{31}+\tau_{41}-\tau_{42})} & 0 \end{bmatrix} \begin{bmatrix} S_1(\omega) \\ S_1(\omega)e^{i\omega(\tau_{21}-\tau_{22}+\tau_{32}-\tau_{31})} \\ S_2(\omega) \\ S_2(\omega)e^{i\omega(\tau_{21}-\tau_{22})} \end{bmatrix}$$

$$= \mathbf{HS} \tag{3.11}$$

Now, separating the two sources from two directions is equivalent to identifying a matrix $\mathbf{W}$, which forms

$$\begin{bmatrix} y_1 \\ y_2 \\ y_3 \\ y_4 \end{bmatrix} = \mathbf{W} \begin{bmatrix} \text{Ch 1} \\ \text{Ch 2} \\ \text{Ch 3} \\ \text{Ch 4} \end{bmatrix}$$

$$= \mathbf{WHS} = \begin{bmatrix} \sigma_1 & 0 & 0 & 0 \\ 0 & \sigma_2 & 0 & 0 \\ 0 & 0 & \sigma_3 & 0 \\ 0 & 0 & 0 & \sigma_4 \end{bmatrix} \begin{bmatrix} S_1(\omega) \\ S_1(\omega)e^{i\omega(\tau_{21}-\tau_{22}+\tau_{32}-\tau_{31})} \\ S_2(\omega) \\ S_2(\omega)e^{i\omega(\tau_{21}-\tau_{22})} \end{bmatrix} \tag{3.12}$$

The contrast function for estimating the unmixing matrix $\mathbf{W}$ is

$$R(\mathbf{W}) = \frac{\int y_1(t) y_2[t - (\tau_{21} - \tau_{22} + \tau_{32} - \tau_{31})]\mathbf{d}t}{\sigma_{y_1}\sigma_{y_2}} \leq 1 \tag{3.13}$$

The equality holds if and only if $y_1(t)$ and $y_2[t - (\tau_{21} - \tau_{22} + \tau_{32} - \tau_{31})]$ are identical signals regardless of the magnitude of ambiguity.

In addition,

$$R(\mathbf{W}) = \frac{\int y_3(t) y_4[t - (\tau_{21} - \tau_{22})]\mathbf{d}t}{\sigma_{y_3}\sigma_{y_4}} \leq 1 \tag{3.14}$$

The equality holds if and only if $y_3(t)$ and $y_4[t-(\tau_{21}-\tau_{22})]$ are identical signals regardless of the magnitude of ambiguity.

Therefore, the source will be separated as long as the inequalities Eq. 3.13 and Eq. 3.14 are maximized. It warrants mention that $\mathbf{H}$ has only six unknown parameters other than $4 \times 4 = 16$ parameters, and this reduces the estimation complexities.

The overall condition number of the array response can now be optimized as a whole. Some proper delays ($\tau_{21}, \tau_{22}, \tau_{32}, \tau_{31}, \tau_{41}$ and τ_{42}) with respect to the microphone-to-microphone distance must be chosen in order to achieve good condition numbers for all frequencies. When $h_{ij} = 1, \tau_{ij} = 0$ for all i, j, the $\mathbf{H}$ is singular whose condition number is infinity. In order to ensure the stability of the inverse, h_{ij} is set to one when optimizing the delays ($\tau_{21}, \tau_{22}, \tau_{32}, \tau_{31}, \tau_{41}$ and τ_{42}).

$$\mathbf{H} = \begin{bmatrix} 1 & 0 & 1 & 0 \\ 1 & 0 & 0 & 1 \\ 0 & 1 & 0 & 1 \\ 0 & 0 & e^{i\omega(\tau_{21}-\tau_{22}+\tau_{32}-\tau_{31}+\tau_{41}-\tau_{42})} & 0 \end{bmatrix} \tag{3.15}$$

Considering a practical scenario where the four microphones are mounted in a straight line along the direction of direct sound and reflected sound, as shown in Fig. 3.1.

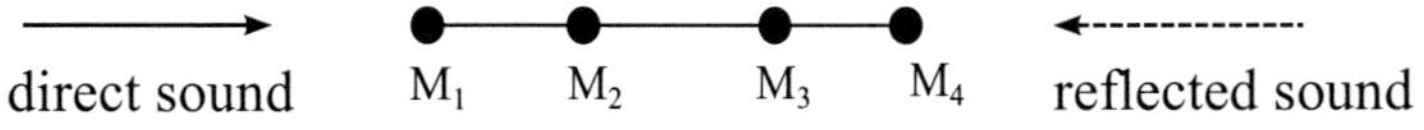

Figure 3.1.: Microphones' positions: Four microphones are mounted in a straight line along the direction of direct sound and reflected sound

The overall length of the array is limited to 50 cm. Two microphones are placed at the beginning and the end of the array. The other two microphones are positioned in the middle. There is no analytical solution for optimizing the condition number, and all the possible positions of the two microphones in the middle are tested. The goal is to find the best microphones' positions where the condition number $\kappa(H)$ is not too large at the frequencies of interest. The microphones' positions are chosen step by step. For example, one microphone's position is selected as $1\text{cm}, 2\text{cm}, \cdots 48\text{cm}$, the other microphone's position is selected as 2cm, 3cm, $\cdots$ 49cm, and the condition number of H of all the possible positions are calculated. Therefore, the optimized microphones' positions are shown in Fig. 3.2. The distances between the microphones are 10cm , 25cm and 15cm. Since the overall length of the array is limited, and this array has to cover the frequency band to the widest extent possible, the microphone permutation must be chosen at different frequencies. M1, M2, M3 and M4 in the first row of Fig. 3.2 are corresponded to Ch1, Ch2, Ch3 and Ch4 respectively. The reciprocal condition numbers $1/\kappa(\mathbf{H})$ at different frequencies are plotted in Fig. 3.3. For example, at 700Hz, if the four microphones from the left to the right are corresponded to the **Ch1, Ch2, Ch3 and Ch4** or **Ch1, Ch4, Ch2 and Ch3**, the mixing matrix **H** is ill-conditioned, and **H** cannot be accurately calculated and estimated. However, if the four microphones correspond to **Ch1, Ch3, Ch4 and Ch2**, the mixing matrix **H** is well-conditioned, and direct sound and reflected sound can be separated correctly. At 200Hz, we have to chose the microphone permutation as **1 2 3 4**. Since the length of this array is limited to 50 cm, the frequency range can cover up to only 3,400 Hz with four microphones. If higher frequencies are required, more microphones should be used. Actually, the physical meaning of this condition number optimization is folding a linear microphone array in a different way.

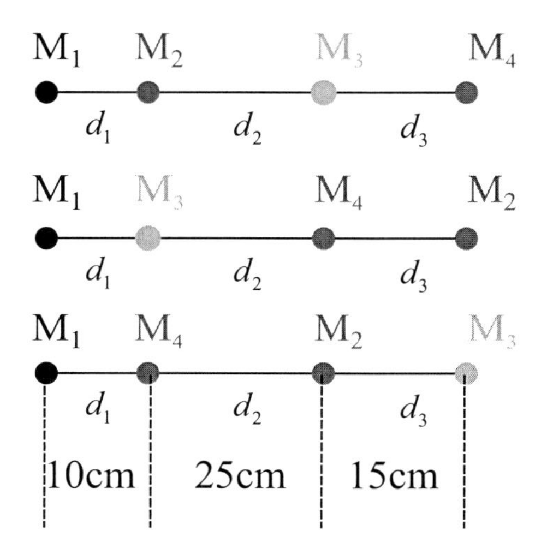

Figure 3.2.: Microphone permutations for broadband signal, different microphone permutations are selected for different frequencies

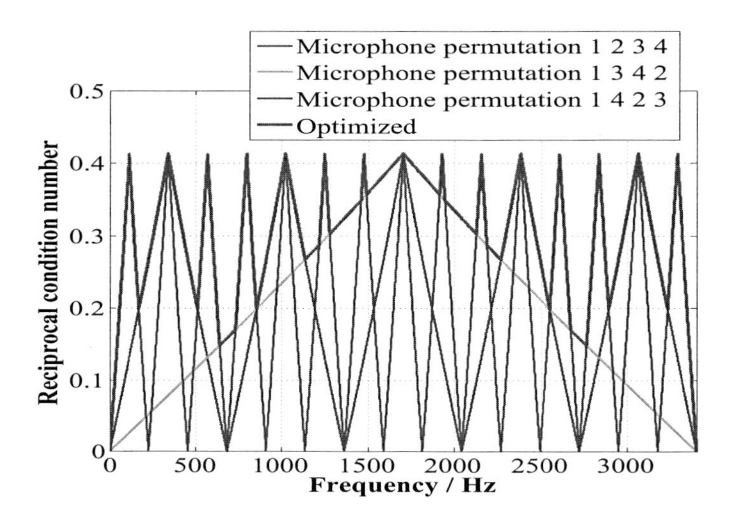

Figure 3.3.: The reciprocal condition number of different microphone permutations corresponding to Fig. 3.2 At some frequencies, the permutation 1234 shows a better condition number than the permutation 1342. The permutation 1234 is selected at these frequencies and vice versa. The optimized permutation selection is denoted by the purple curve.

The correlation contrast function Eq. 3.13 ignores the absolute magnitude of signals. After the direct sound and reflection sound is separated, one more step has to be carried out to obtain the absolute value of the signal.

Assuming that s_{est1} and s_{est2} are the separated direct sound and reflection sound, the magnitude of each signal can be estimated by a least mean square error curve fitting for the parameters α_1 and α_2.

$$\text{Ch } 1 = \alpha_1 \cdot s_{\text{est1}}(t) + \alpha_2 \cdot s_{\text{est2}}(t) \tag{3.16}$$

3.3. Application:reflection coefficient *in situ* measurements

A possible application for the multi-microphone technique discussed above is measuring the reflection coefficient. The standard approach to measure the

Figure 3.4.: Sound barrier measurement. The microphones are placed in front of the wall and a loudspeaker is mounted approximately 1.5 m away from the wall.

impedance or the corresponding reflection coefficient and absorption coefficient is using Kundt's tube [AST] [ISO, 1998]or the reverberation chamber [ISO, 2003].

Under standard laboratory conditions, the measurements are indeed accurate. However, in various cases of monitoring the acoustic performance of surfaces or in acoustic design, the reflection coefficient has to be measured in situ as well.

Numerous approaches have been developed by several authors. The measurement approach for the normal incident reflection coefficient is illustrated by [Garai, 1993] and [Mommertz, 1995]. The oblique incidence approach is described by [Yuzawa, 1975] [Mulholland, 1979] [Nocke, 2000]. An alternative way to measure the acoustic impedance is using the pu-probe, which directly measures the particle velocity of the air vibration [Lanoye, 2006]. The impedance is then obtained directly from the ratio between the p and u outputs.

The normal-incidence approach is discussed in greater detail here. The loudspeaker and the microphone are positioned in front of the surface under test (see Fig. 3.4). The sound wave is normally incident on the surface, and the reflection coefficient can be obtained by analyzing the reflected sound.

As illustrated in Fig.3.5, the impulse response of this system contains a direct sound pulse, a reflected sound pulse from the surface under test, and several other reflections from the ground or nearby objects. In the time domain, the reflections from the ground and other objects can be easily windowed out.

In case the microphone is positioned very close to the surface under test, the direct sound pulse is very close to the reflected sound pulse, and there is some overlap of pulses, especially at low frequencies. In addition, some information is lost when the time window is implemented. [Garai, 1993]. Then an alternative subtraction method is implemented by [Mommertz, 1995].

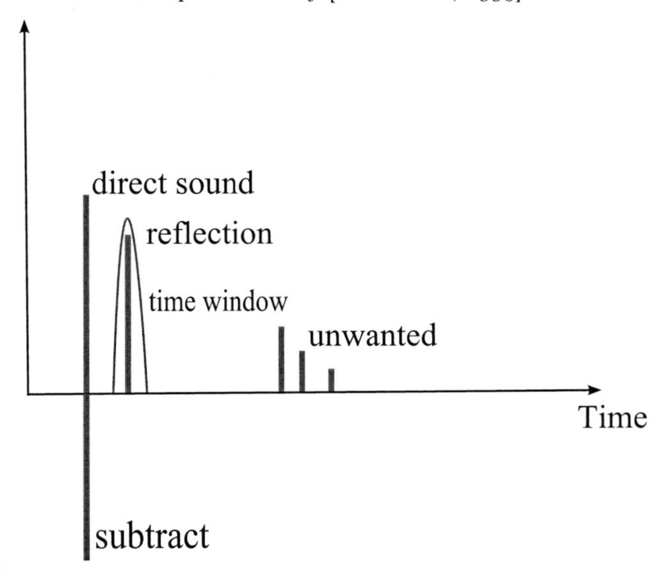

Figure 3.5.: Impulse response in a reflection factor measurement. There are two ways to cancel the direct sound: multiplying a time window on the reflected sound or subtracting the direct sound obtained without the sound barrier. If the microphone is far from the surface under test, time windowing can be implemented. If the microphone is very close to sound barrier, the subtraction method should be used.

In the subtraction method, a reference sound is first measured in free-field conditions. Then the loudspeaker and the microphone are positioned in front of the surface under test with an identical relative position. Afterwards, the overall impulse response, which contains the both the direct sound and the reflected sound, is measured. In this case, the direct sound, which is included in the overall impulse response, should be identical to the reference direct sound which is measured in the free-field room. The reflected sound can be extracted by subtracting the reference direct sound from the overall impulse response. The advantage of the subtraction method is that no time windowing is needed. The drawback is that it is highly susceptible to variations of the measurement setup and environmental conditions, such as temperature changes, air movement, and imperfect positioning of the microphones. Those variations may cause temporal misalignments of the direct sound, which will lead to subtraction errors, especially at high frequencies. In order to align the delay, [Xiang, 2010] arbitrarily compensates for the time misalignment by estimating the delay between the reference measurement and the in situ measurement. In recent years, an inverse filter approach is being applied to equalize the impulse response of the loudspeaker [Wehr, 2013] and make the loudspeaker's impulse response as short as possible. The overlap area of the impulse response between direct sound and reflected sound can be reduced and the time-windowing can be performed; thereby, the reference measurement in a free-field condition for subtraction is avoided.

In order to avoid the use of reference sound measurement in free-field conditions, the aforementioned microphone array technique can be used as well. In the microphone array, the reflection coefficient can be obtained through a single measurement. It is not necessary to move the devices to the free-field room and measure the reference direct sound once again.

3.3.1. Measurement results

A measurement is carried out in an anechoic chamber to investigate the applicability of separating direct sound and reflected sound for reflection coefficient measurement using the array aperture described in Fig. 3.2.

The microphones have to be broadband calibrated and synchronized. Otherwise, even a minor misalignment of signals will lead to errors. The overall impulse response contains two impulses: one belongs to the direct sound, the other belongs to the reflected sound, and they are very close to each other. The red curve in Fig. 3.7 and Fig. 3.8 denotes the measurement results by the subtraction method. Since the subtraction method is performed in a free-field room, all the relative positions of the microphones and loudspeakers are kept perfectly constant, and the results of the subtraction methods can be used as a reference to evaluate the performance of the four-microphones approach and the time-windowing approach. If time windowing is performed to window out the direct sound, large errors occur at low frequencies—e.g. around 10 dB at 330 Hz (Fig. 3.8)—because the overlap area between the direct sound component and the reflected sound component is larger at low frequencies. If the reflected sound is separated using four microphones, the transfer functions are very close to the transfer function resulting from the subtraction methods. At high frequencies of over 1000 Hz, the three different methods show the same results.

Fig. 3.9 shows the necessity of optimizing the condition number. If the condition number of the array is not optimized as per Fig. 3.3, at the frequencies where the condition numbers are too large, the estimation errors will be as large as 10 dB.

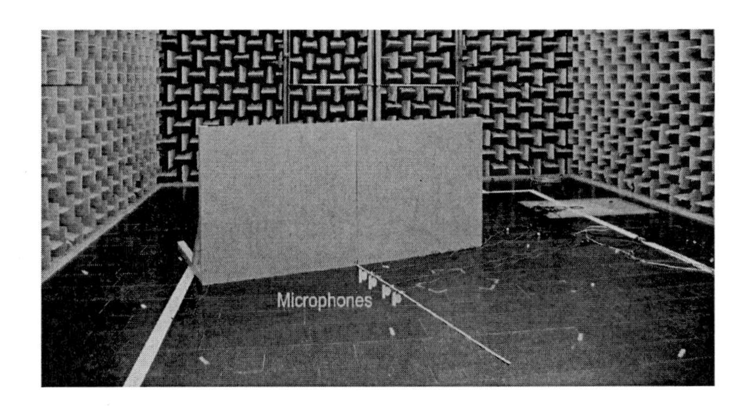

Figure 3.6.: An example to separate reflected sound and direct sound. A wood wall is positioned in the anechoic chamber. Four microphones are placed on the floor in a straight line.

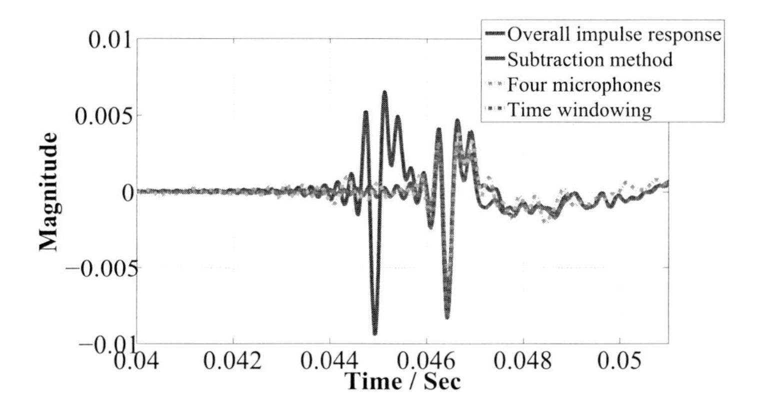

Figure 3.7.: The impulse responses obtained by different methods. The overall impulse response contains a direct sound pulse and a reflected sound pulse. The direct sound component is canceled by three different methods: subtraction method, the four-microphone array and time windowing

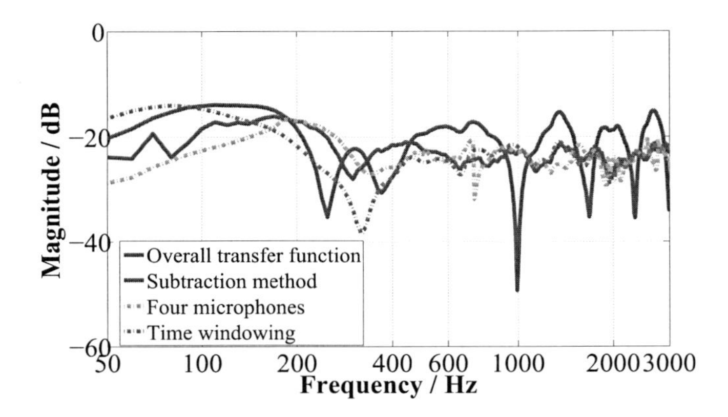

Figure 3.8.: The transfer functions of the reflected sound obtained by different methods. The measurement is performed in the free-field room. The direct sound is measured first. Then a wood wall is placed in front of the microphones without moving the microphones and the loudspeakers. Thus, the direct sound can be considered to be perfectly subtracted by the subtraction method. Hence, the red curve for the subtraction method can be used as a reference to compare the time-windowing method and the four-microphone array method.

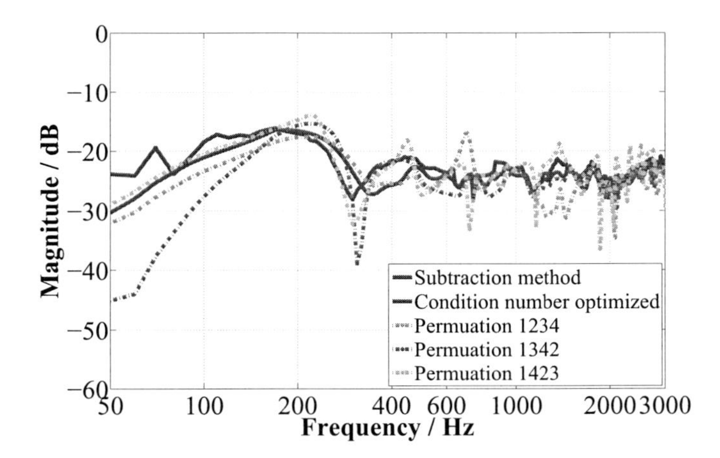

Figure 3.9.: Comparison of the array optimization. If the condition numbers of the array are not optimized, at the frequencies where the condition numbers are too large, the estimation errors will be large.

3.4. Discussion

A four-microphone array can be applied to separate direct sound and reflected sound in reflection coefficient measurements. However, actual implementations of this approach have their limitations.

Firstly, the underlying assumption of this method is that the attenuation coefficients between different microphones are constant at different frequencies. In other words, the mixing matrix **H** in Eq. 3.11 remains identical over all frequencies. To satisfy this requirement, the line of the four microphones has to be perpendicular to the reflection surface under test. At the same time, the surface under test cannot have very complicated structures of scattering. This limits the practical applications.

Secondly, if one noise is active, more microphones are required to estimate and suppress the noise. Assuming that the noise consists of only direct sound from a different direction, six microphones will be required to estimate the magnitude of the sound from three directions. This assumption is made because if the method does not work in such simple case, it cannot work in more complicated scenarios.

Similar to Eq. 3.4 and Eq. 3.5, the formulation to separate the sounds from three different directions is represented by the equations Eq. 3.17 to Eq. 3.21. To recover the sound from three different directions, at least three microphones are needed, as shown in Eq. 3.17. $S_d(\omega)$ is the direct sound, $S_r(\omega)$ is the reflected sound and $N(\omega)$ is the direct sound of noise.

$$\begin{aligned}\mathsf{X}_{123} = \begin{bmatrix} X_1(\omega) \\ X_2(\omega) \\ X_3(\omega) \end{bmatrix} &= \begin{bmatrix} h_{11}e^{-i\omega\tau_{11}} & h_{12}e^{-i\omega\tau_{12}} & h_{43}e^{-i\omega\tau_{13}} \\ h_{21}e^{-i\omega\tau_{21}} & h_{22}e^{-i\omega\tau_{22}} & h_{52}e^{-i\omega\tau_{23}} \\ h_{31}e^{-i\omega\tau_{31}} & h_{32}e^{-i\omega\tau_{32}} & h_{52}e^{-i\omega\tau_{33}} \end{bmatrix} \begin{bmatrix} S_d(\omega) \\ S_r(\omega) \\ N(\omega) \end{bmatrix} \\ &= \mathsf{H}_{123}\mathbf{S} \end{aligned} \tag{3.17}$$

Even if the angles of the desired source and noise are known exactly, the mixing matrix H_{123} is still unknown. Therefore, three additional microphones are used to estimate the magnitudes $h_{ij}, i = 1, 2, ...6, j = 1, 2, 3$.

$$\mathsf{X}_{456} = \begin{bmatrix} X_4(\omega) \\ X_5(\omega) \\ X_6(\omega) \end{bmatrix} = \begin{bmatrix} h_{41}e^{-i\omega\tau_{41}} & h_{42}e^{-i\omega\tau_{42}} & h_{43}e^{-i\omega\tau_{43}} \\ h_{51}e^{-i\omega\tau_{51}} & h_{52}e^{-i\omega\tau_{52}} & h_{53}e^{-i\omega\tau_{53}} \\ h_{61}e^{-i\omega\tau_{61}} & h_{62}e^{-i\omega\tau_{62}} & h_{63}e^{-i\omega\tau_{63}} \end{bmatrix} \begin{bmatrix} S_d(\omega) \\ S_r(\omega) \\ N(\omega) \end{bmatrix}$$
$$= \mathsf{H}_{456}\mathbf{S} \tag{3.18}$$

If two unmixing matrices $\mathbf{W}_{123}$ and $\mathbf{W}_{456}$ are found, which hold true for Eq. 3.19 and Eq. 3.20, the direct sound, reflected sound and noise are separated.

$$\mathsf{W}_{123}\mathsf{H}_{123}\mathbf{S} = \begin{bmatrix} \mathsf{y}_1 \\ \mathsf{y}_2 \\ \mathsf{y}_3 \end{bmatrix} = \mathbf{C} \begin{bmatrix} S_d(\omega) \\ S_r(\omega) \\ N(\omega) \end{bmatrix}$$
$$\text{where } \mathbf{C} = \begin{bmatrix} c_1 & 0 & 0 \\ 0 & c_2 & 0 \\ 0 & 0 & c_3 \end{bmatrix} \tag{3.19}$$

$$\mathsf{W}_{456}\mathsf{H}_{456}\mathbf{S} = \begin{bmatrix} \mathsf{y}_4 \\ \mathsf{y}_5 \\ \mathsf{y}_6 \end{bmatrix} = \mathbf{C} \begin{bmatrix} S_d(\omega) \\ S_r(\omega) \\ N(\omega) \end{bmatrix}$$
$$\text{where } \mathbf{C} = \begin{bmatrix} c_4 & 0 & 0 \\ 0 & c_5 & 0 \\ 0 & 0 & c_6 \end{bmatrix} \tag{3.20}$$

The contrast function for estimating the magnitude matrices $\mathbf{H}_{123}$ and $\mathbf{H}_{456}$ is the correlation to Eq. 3.21

$$Max\left\{\text{Corr}\left(\mathsf{y}_1, \mathsf{y}_4\right) \cdot \text{Corr}\left(\mathsf{y}_2, \mathsf{y}_5\right) \cdot \text{Corr}\left(\mathsf{y}_3, \mathsf{y}_6\right)\right\} \tag{3.21}$$

But this procedure requires too many microphones. When the scattering or reverberation effect needs to be measured, the separation of the noise becomes impossible. Therefore, the applicability of the BSS to separate the noise on the impulse response measurement is investigated in the next chapter.

4 Applicability of source separation method on impulse response measurement

In the example of the impulse response measurement outdoors, the measurement may have to be performed under severe conditions or a stationary background noise. In this case, averaging is a conventional and robust way to improve the SNR. If the signal is averaged N times, the SNR will be increased by $10 \times \log_{10} N$ dB as long as the system under test is a time-invariant system. But long term averaging suffers from the risk of time variances. Therefore, this chapter aims to investigate the possibilities that might separate the desired signal from the noise without averaging.

An intuitive solution is the beamforming technique (also called 'spatial filtering'). Chapter 3 explains that direct sound and reflected sound can be separated by a modified spatial filtering where the magnitude of each channel is estimated by correlation. The same technique can be applied to separate the two sources. If two sources are located in a free field, four microphones are required to estimate the magnitudes arriving at different microphones. If the sources have a couple of reflections, more microphones are required. If the impulse response of a reverberant system is measured in the presence of noise, the spatial filter fails because too many unknown parameters have to be estimated. BSS provides an alternative way to construct contrast functions and separate the signals by utilizing the properties of statistical independence between sources.

This chapter illustrates four examples to show that a BSS-based statistical method cannot be applied for the impulse response measurement as well. The first example involves the separation of two sources in a free field and both of the two sources have only direct sounds but no reflections. Two sensors are used to separate the sources. This is the fundamental model of BSS. In the second example, one source has one direct sound and one reflection, while the other source has only direct sound. Three sensors are used to separate the sources. The third example shows that if both sources have a few reflections, the source separation method fails. In the fourth example, the impulse response is measured using sweep in a reverberant environment. Since too many parameters have to be estimated in order to separate the sweep and the noise, it is impossible to estimate all the parameters through a single measurement. Multiple sweeps are used to construct independent signal sequences and contrast functions. This can separate the desired signal from the noise, but compared with the averaging method, the source separation method shows no advantage.

4.1. Separate two waves in a free field

Two sources are located in a free field and have only direct sounds, as shown in Fig. 4.1. If the sound waves are assumed to be plane waves, then conventional beamforming can be applied. Chapter 3 illustrates a particular case where a four-microphone method is used to separate direct sound and reflected sound. The same approach can also be used to separate two independent sources in a free field.

Nevertheless, if joint diagonalization (2.63) is used as the contrast function in this scenario, only two microphones are needed.

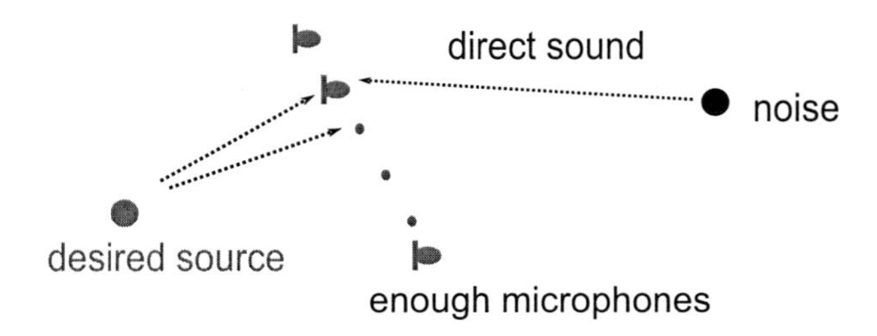

Figure 4.1.: Two sources are located in a free field, and they have only direct sounds

The impulse response is simulated as in Fig. 4.2. The impulse responses from the sources to the microphones are only Dirac impulses with different delays and magnitudes. The delays can be obtained from the known locations of the sources. If the locations of the sources are unknown, the delays can also be estimated using DOA [Dudgeon, 1993]. In any case, the delays here are assumed to be known. Only the magnitudes of the impulses that are the $h_{ij}, i = 1, 2, j = 1, 2$ in Eq. 4.1 should be estimated.

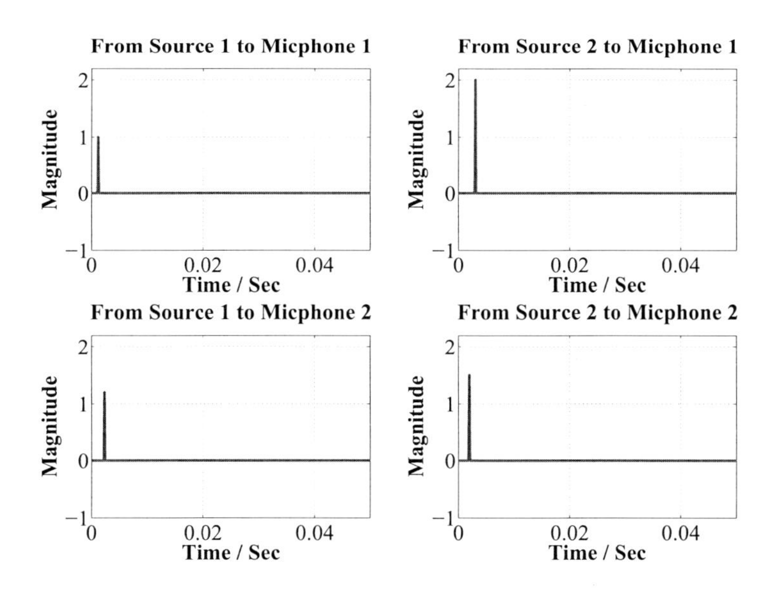

Figure 4.2.: The impulse response from two sources to two microphones. The impulse responses are Dirac impulses with different delays and magnitudes.

$$\begin{aligned} \text{Mic 1}: \quad & x_1(t) = h_{11}s_1(t - \tau_{11}) \quad && + \quad h_{12}s_2(t - \tau_{12}) \\ \text{Mic 2}: \quad & x_2(t) = h_{21}s_1(t - \tau_{21}) \quad && + \quad h_{22}s_2(t - \tau_{22}) \end{aligned} \tag{4.1}$$

The separation procedure is performed as follows.

Firstly, Eq. 4.1 is transformed to the frequency domain.

$$
\begin{bmatrix} X_1(\omega) \\ X_2(\omega) \end{bmatrix} = \begin{bmatrix} h_{11}e^{-i\omega\tau_{11}} & h_{12}e^{-i\omega\tau_{12}} \\ h_{21}e^{-i\omega\tau_{21}} & h_{22}e^{-i\omega\tau_{22}} \end{bmatrix} \begin{bmatrix} S_1(\omega) \\ S_2(\omega) \end{bmatrix}
$$
$$
\mathbf{X} = \mathbf{HS} \tag{4.2}
$$

The second step involves calculating the symbolic inverse of $\mathbf{H}$ in Eq. 4.2, which is notated as $\mathbf{W} = \mathbf{H}^{-1}$. τ_{11}, τ_{12}, τ_{21} and τ_{22} are assumed to be known, while $\mathbf{W}$ is a function with respect to $h_{ij}(i = 1,2, j = 1,2)$ and is written as $\mathbf{W}(h_{11}, h_{12}, h_{21}, h_{22})$. This step reduces the computing complexity for estimating the unmixing matrix.

The third step consists of choosing an initial h_{11}, h_{12}, h_{21} and h_{22} randomly, and computing $\mathbf{Z} = \mathbf{WX}$ in the frequency domain. Calculate and minimize the contrast function Eq. 2.63 in the time domain.

The non-Gaussianity function Eq. 2.44 does not work in this case. The separation results are shown in Fig. 4.3. The first row represents the initial sweep and noise. They are convolved with the impulse responses shown in Fig. 4.2. The mixed signals are shown in the second row. The separation results are shown in the third row. Since only four parameters $h_{ij}, i = 1,2, j = 1,2$ are estimated, the sweep is correctly separated.

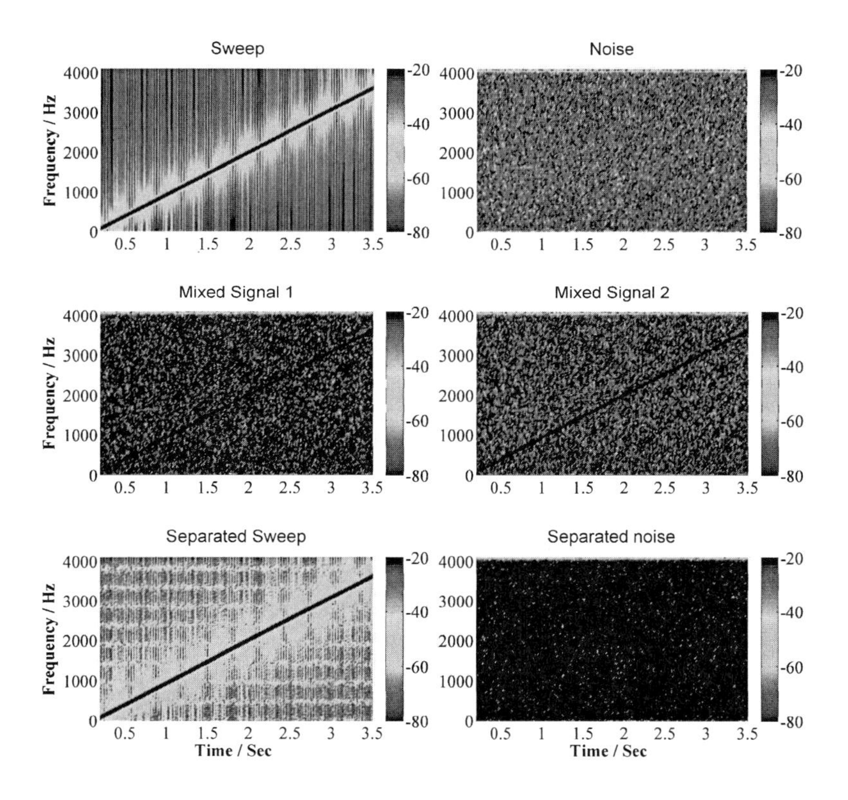

Figure 4.3.: The performance of source separation using cross-correlation as the contrast function. The first row is the initial sweep and noise. The second row is the mixed signal. The third row is the separated noise and sweep. The sweep is separated.

4.2. Separating direct and reflected waves and noise

In the second example, one source has one direct sound and one reflection, whereas the other source has only direct sound. A possible scenario is the sound barrier measurement and it involves a noise source at the corner of the barrier(Fig. 4.4). This scenario is investigated because if BSS cannot separate the source signal and the noise signal in such a simple scenario, it cannot separate the sources in more complicated scenarios.

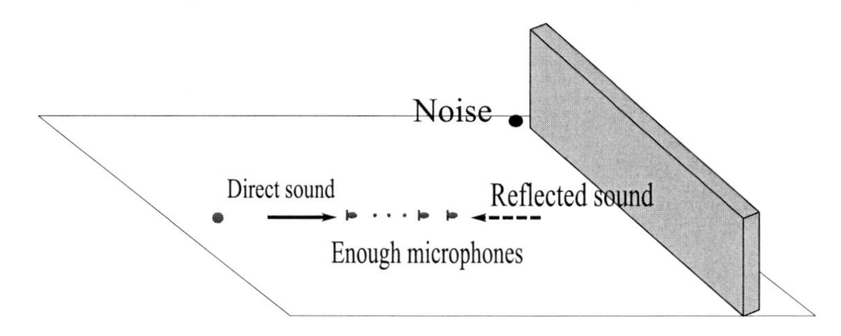

Figure 4.4.: The desired source has direct sound and reflected sound. The noise source has only direct sound.

In order to separate the sound from three different directions mathematically, at least three microphones should be used. The mixture equations are re-written as Fig.4.3 for convenience.

$$\mathsf{X}_{123} = \begin{bmatrix} X_1(\omega) \\ X_2(\omega) \\ X_3(\omega) \end{bmatrix} = \begin{bmatrix} h_{11}e^{-i\omega\tau_{11}} & h_{12}e^{-i\omega\tau_{12}} & h_{13}e^{-i\omega\tau_{13}} \\ h_{21}e^{-i\omega\tau_{21}} & h_{22}e^{-i\omega\tau_{22}} & h_{23}e^{-i\omega\tau_{23}} \\ h_{31}e^{-i\omega\tau_{31}} & h_{32}e^{-i\omega\tau_{32}} & h_{33}e^{-i\omega\tau_{33}} \end{bmatrix} \begin{bmatrix} S_d(\omega) \\ S_r(\omega) \\ N(\omega) \end{bmatrix} \tag{4.3}$$

where $S_d(\omega)$ is the direct sound from the loudspeaker and $S_r(\omega)$ is the reflected sound from the wall. $S_d(\omega)$ and $S_r(\omega)$ are correlated. $N(\omega)$ is the direct sound of the noise. The excitation signal is a linear sweep and the noise is uncorrelated with the source. The delays $\tau_{i,j}, i, j = 1, 2, 3$ are assumed to be known. The magnitudes $h_{ij}, i, j = 1, 2, 3$ are unknown and have to be estimated. The joint diagonalization (Eq. 2.63) is used as the contrast function. The contrast function of non-Gaussianity does not work in this scenario either.

The simulated impulse responses from the sources to the microphones are illustrated in Fig. 4.5. The impulse responses from the loudspeaker to the microphones include two impulses. The first impulse is the direct sound from the loudspeaker and the second impulse is the reflected sound from the wall. The only differences between the impulse responses of the three microphones are the magnitudes and the delays. The delays of the direct sound, reflected sound and the noise are assumed to be a priori known parameters.

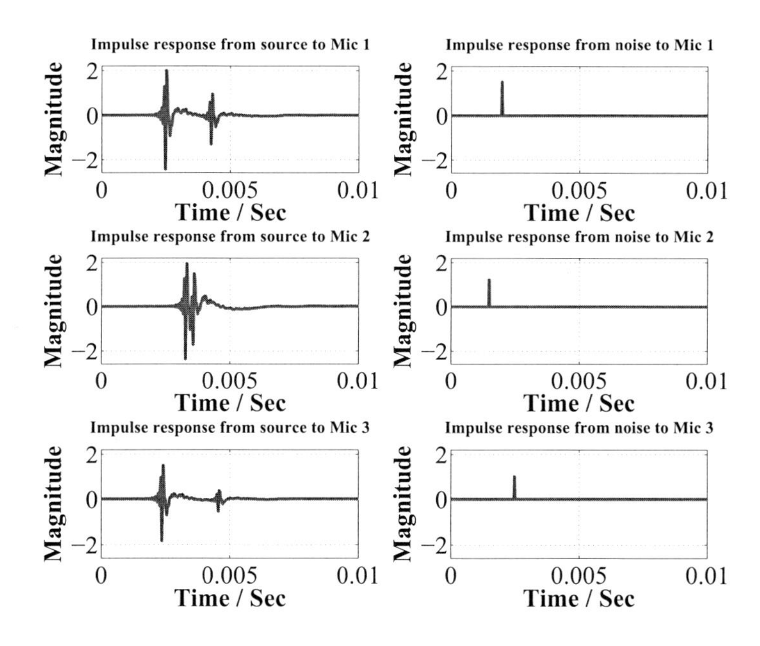

Figure 4.5.: The simulated impulse response from the sources to the receivers

The separation results are shown in Fig.4.6. The first row shows the original sweep and the noise. The second row illustrates the mixed signals recorded by the three microphones. The noise is so strong that the sweep is masked. The separation results are presented in the third row. The sweep appears to be separated from the noise because the separated signal has a sweep-like spectrogram, but this sweep is not the original sweep but only a filtered version. Since the contrast function of joint diagonalization Eq. 2.63 cannot distinguish between the correlated signals and since direct sound is correlated with the reflected sound, direct sound and reflected sound cannot be separated.

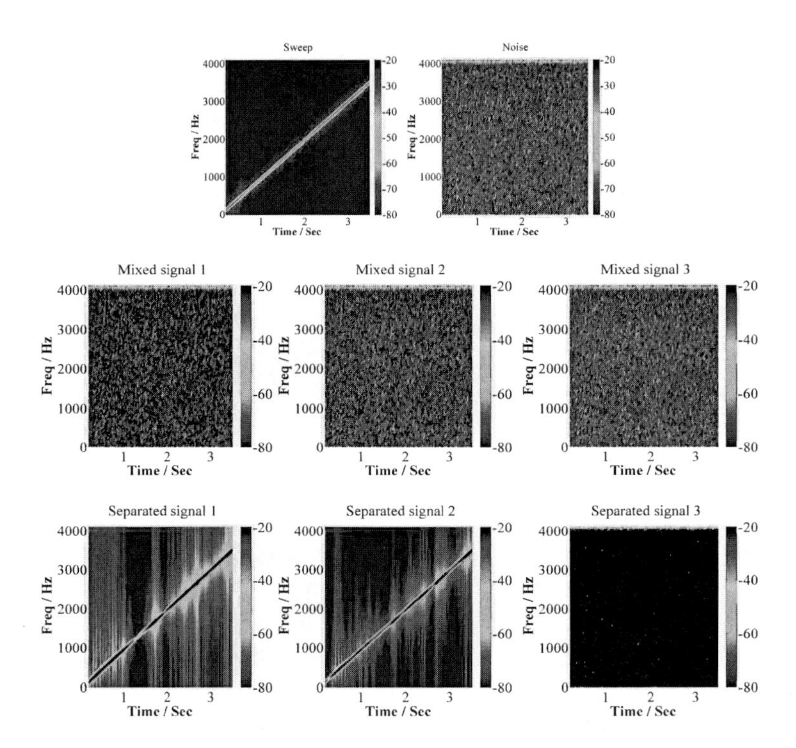

Figure 4.6.: The spectrogram of the simulated original sweep and noise, the mixed signals and the separated signals. After the separation, the noise is canceled. But the sweep is not the original sweep; it is a filtered version of the original sweep

There are nine unknown parameters, $h_{ij}, i = 1, 2, 3, j = 1, 2, 3$, in Eq. 4.3, but only one contrast function, which is joint diagonalization. The solutions are not unique. The contrast functions have more than one maximum, as seen in Fig. 4.7. Since Eq.4.3 is still ill-posed, the joint diagonalization can maximize the independence between the two sources, and the noise is separated accordingly. But the solutions are not unique. If different initial points for the iteration are chosen, the iteration will converge to different local maxima and the final solutions will differ, as shown in Fig. 4.8. The noise is separated, but the source signal is distorted.

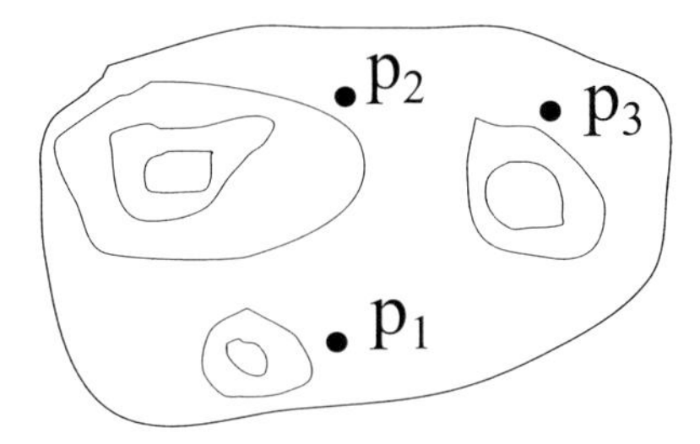

Figure 4.7.: The contrast function with local minima. The contrast function has more than one local minimum. If different initial points are chosen, the contrast function will converge to different local minima.

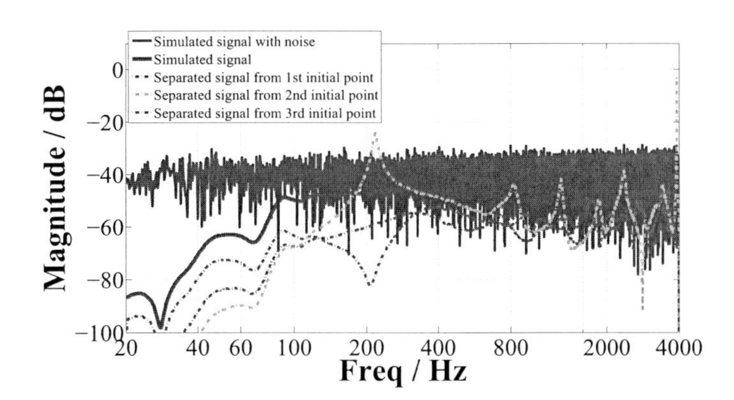

Figure 4.8.: The spectrum of the separated sweep signals corresponding to Fig. 4.6 through different initial iteration points. The initial points are chosen randomly. The solutions are not unique. The contrast function of joint diagonalization can maximize the independence between the two sources. The noise is separated, but the source signal is distorted.

4.3. Each source has a few reflections

Considering a slightly more practical scenario, both sources have a few reflections (Fig. 4.9). The delays of the reflections are unknown. This is a MIMO inverse problem. Since too many parameters are unknown, the time domain procedure (Eq. 2.65 to Eq. 2.68) is used instead of the symbolic matrix inverse in Sec. 4.1 and Sec. 4.2.

The BSS is arbitrarily implemented on the mixed signal. The sampling rate is 8,192 Hz and the order of the unmixing FIR filer is chosen as 160 orders. There is only one contrast function but 160 * 4 unknown parameters. The separation results are not unique. One separation result is shown in Fig. 4.10. It is difficult to define the improvement of the SNR, and the sweep is anyway distorted.

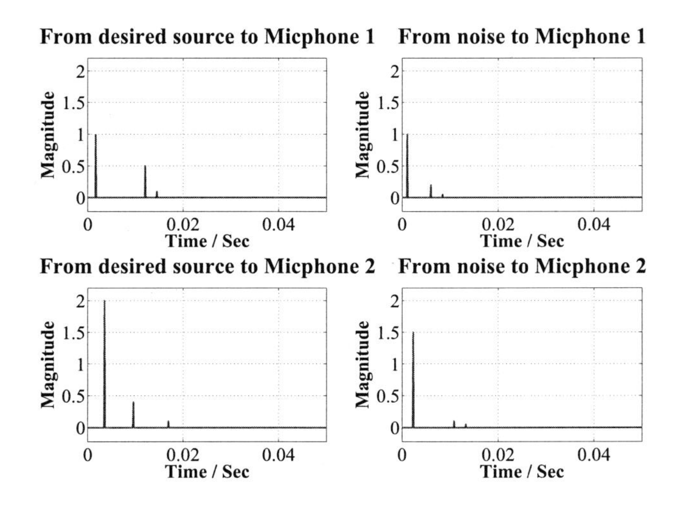

Figure 4.9.: The simulated impulse responses from the desired source and noise to the two microphones. There are three reflections in each of the impulse responses.

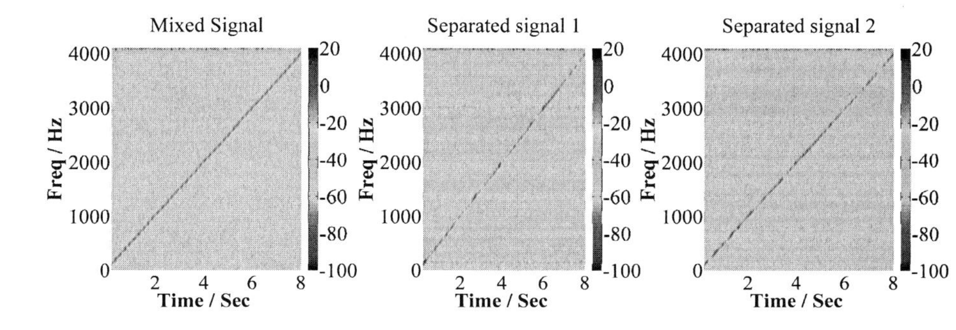

Figure 4.10.: The mixed signal and separated signals. It is difficult to define the SNR improvement of the sweep and the sweep is distorted.

4.4. Separation of noise in reverberant fields

If the number of reflections increases, or the scattering and reverberation have to be measured, the number of unknown parameters increases rapidly. For example, if the reverberation time of the measured system is one second and the sampling rate is 8,192 Hz, then one desired source and one noise source are present. Two microphones are used to separate the noise and the separation is performed in the frequency domain. The mixing matrix is a 2×2 matrix for every frequency bin. There are 4097 frequency bins which means $2 \times 2 \times 4{,}097$ unknown parameters should be estimated simultaneously. It becomes impossible to estimate all those parameters by a single measurement using only one assumption of independence. Another way to separate the noise is by using multiple sweeps in order to get sufficient samples. The equation for measured signals is also written as 4.4,

$$\begin{bmatrix} X_1(\omega) \\ X_2(\omega) \end{bmatrix} = \begin{bmatrix} H_{11}(\omega) & H_{12}(\omega) \\ H_{21}(\omega) & H_{22}(\omega) \end{bmatrix} \begin{bmatrix} S(\omega) \\ N(\omega) \end{bmatrix} \tag{4.4}$$

where $X_1(\omega), X_2(\omega)$ are the recorded signals in the frequency domain. $S(\omega)$ is the excitation signal and $N(\omega)$ is the noise. $H_{ij}(\omega)$ is the transfer function from the source and the noise to the sensors. If there is no noise, the transfer functions from the source to the sensors, $H_{11}(\omega)$ and $H_{21}(\omega)$, are directly measured.

Since the noise $N(\omega)$ appears and is unknown, the transfer functions cannot be measured directly. This is an ill-posed inverse problem because the number of unknown parameters $H_{ij}(\omega), i = 1, 2, j = 1, 2$ and $N(\omega)$ are more than the number equations. However, these equations can still be solved if the measurements are repeated and the sequences of repeated excitation signals are independent from the noise. This solution is formulated as Eq. 4.5.

$$\begin{bmatrix} X_1(\omega) \\ X_2(\omega) \end{bmatrix} = \begin{bmatrix} H_{11}(\omega) & H_{12}(\omega) \\ H_{21}(\omega) & H_{22}(\omega) \end{bmatrix} \begin{bmatrix} S_1(\omega) & S_2(\omega) & \cdots & S_m(\omega) \\ N_1(\omega) & N_2(\omega) & \cdots & N_m(\omega) \end{bmatrix} \tag{4.5}$$

where $S_m(\omega)$ is the m-th excitation signal, and $N_m(\omega)$ is noise during the m-th measurement.

If the source sequence $\begin{bmatrix} S_1(\omega) & S_2(\omega) & \cdots & S_m(\omega) \end{bmatrix}$ is uncorrelated with the noise sequence $\begin{bmatrix} N_1(\omega) & N_2(\omega) & \cdots & N_m(\omega) \end{bmatrix}$, then, theoretically, the source and the noise can be separated from the observations $X_1(\omega)$ and $X_2(\omega)$. Fortunately, the source sequence $\begin{bmatrix} S_1(\omega) & S_2(\omega) & \cdots & S_m(\omega) \end{bmatrix}$ is fully known as the excitation signal and can be freely manipulated. We can thus design a sequence that is uncorrelated with the noise sequence. In case of repeated sweeps, these can be designed to be nearly identical, the only difference between them being their polarity. For example, the first three segments are sweeps with the sign +1, and the fourth, sixth and eighth segments are the initial sweeps multiplied by -1. Actually, this +1 and -1 sequence is a MLS sequence. Hence, the excitation signal is built using a sequence of sweep repetition modulated by a MLS sequence. Note that some zeros have to be padded between the sweeps to measure the decay tails of the impulse response [Müller and Massarani, 2001].

This kind of excitation signal is sent to the system under test. The Fourier transform is performed on one segment at a time and segment by segment. The measured signal can be represented by Eq.4.6.

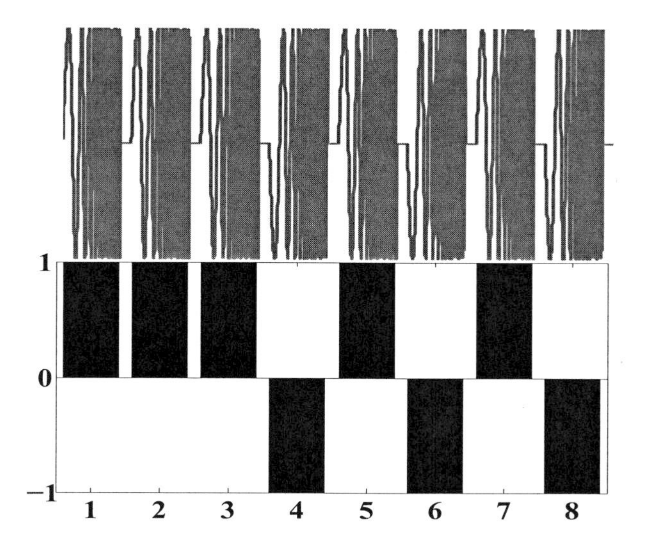

Figure 4.11.: Sweep repetition modulated by MLS

$$\begin{bmatrix} X_{11}(\omega) & X_{12}(\omega) & \cdots & X_{1m}(\omega) \\ X_{21}(\omega) & X_{22}(\omega) & \cdots & X_{2m}(\omega) \end{bmatrix} = \begin{bmatrix} H_{11}(\omega) & H_{12}(\omega) \\ H_{21}(\omega) & H_{22}(\omega) \end{bmatrix} \times \begin{bmatrix} S_1(\omega) & 0 \\ 0 & 1 \end{bmatrix} \begin{bmatrix} \xi_1 & \xi_2 & \cdots & \xi_m \\ N_1(\omega) & N_2(\omega) & \cdots & N_m(\omega) \end{bmatrix} \tag{4.6}$$

where $S_1(\omega)$ is the initial sweep, $\begin{bmatrix} \xi_1 & \xi_2 & \cdots & \xi_m \end{bmatrix}$ is the MLS sequence with the length m, $\begin{bmatrix} N_1(\omega) & N_2(\omega) & \cdots & N_m(\omega) \end{bmatrix}$ is the noise of m measurements which is unknown.

The Eq. 4.6 is simplified as

$$\begin{bmatrix} X_{11}(\omega) & X_{12}(\omega) & \cdots & X_{1m}(\omega) \\ X_{21}(\omega) & X_{22}(\omega) & \cdots & X_{2m}(\omega) \end{bmatrix} = \begin{bmatrix} H_{11}(\omega)S_1(\omega) & H_{12}(\omega)N_1(\omega) \\ H_{21}(\omega)S_1(\omega) & H_{22}(\omega)N_1(\omega) \end{bmatrix} \times \begin{bmatrix} \xi_1 & \xi_2 & \cdots & \xi_m \\ \eta_1 & \eta_2 & \cdots & \eta_m \end{bmatrix}$$

simply notated as

$$\mathbf{X} = \mathbf{B} \begin{bmatrix} & \mathsf{MLS} & & \\ \eta_1 & \eta_2 & \cdots & \eta_m \end{bmatrix} \tag{4.7}$$

where $\eta_m = \frac{N_m(\omega)}{N_1(\omega)}$. Since the MLS sequence is uncorrelated with all other sequences, the excitation signal and the noise is separable.

Afterwards, the noise sequence $\begin{bmatrix} \eta_1 & \eta_2 & \cdots & \eta_m \end{bmatrix}$ is notated as z_2. Since the source sequence $\begin{bmatrix} \xi_1 & \xi_2 & \cdots & \xi_m \end{bmatrix}$ is known by the excitation signal and is the MLS sequence. The mixture equation is rewritten as

$$\mathbf{X} = \begin{bmatrix} H_{11}(\omega)S_1(\omega) & H_{12}(\omega)N_1(\omega)/\alpha_2 \\ H_{21}(\omega)S_1(\omega) & H_{22}(\omega)N_1(\omega)/\alpha_2 \end{bmatrix} \begin{bmatrix} \mathsf{MLS} \\ \mathsf{z}_2 \end{bmatrix} \tag{4.8}$$

Since the noise sequence z_2 is unknown, z_2 must first be separated using the ICA method. Joint diagonalization can also be used as the contrast function for the estimation of z_2, and it will give similar results. Then the mixing matrix $\begin{bmatrix} H_{11}(\omega)S_1(\omega) & H_{12}(\omega)N_1(\omega)/\alpha_2 \\ H_{21}(\omega)S_1(\omega) & H_{22}(\omega)N_1(\omega)/\alpha_2 \end{bmatrix}$ of Eq. 4.8 is obtained by Eq. 4.9

$$\begin{bmatrix} H_{11}(\omega)S_1(\omega) & H_{12}(\omega)N_1(\omega)/\alpha_2 \\ H_{21}(\omega)S_1(\omega) & H_{22}(\omega)N_1(\omega)/\alpha_2 \end{bmatrix} = \mathbf{X} \begin{bmatrix} \mathsf{MLS} \\ \mathsf{z}_2 \end{bmatrix}^+ \tag{4.9}$$

'+' stands for the Moore–Penrose pseudoinverse. The first element of the matrix $\mathbf{X} \begin{bmatrix} \mathsf{MLS} \\ \mathsf{z}_2 \end{bmatrix}$ is the desired signal, $H_{11}(\omega)S_1(\omega)$, without the noise recorded by the first microphone. Although the absolute magnitude of the noise is still unknown, the absolute magnitude of the measured signal without the noise has already been extracted.

A simulation is carried out to evaluate the applicability of this source separation method on the transfer function measurement. The simulated transfer functions are presented in Fig.4.12. The simulated transfer functions are convolved with the MLS-modulated repeated sweeps of Fig. 4.11 and the white noise source. The mixed signals are recorded by two microphones. The complex-valued ICA approach is performed in the frequency domain and the mixing matrix is estimated frequency by frequency.

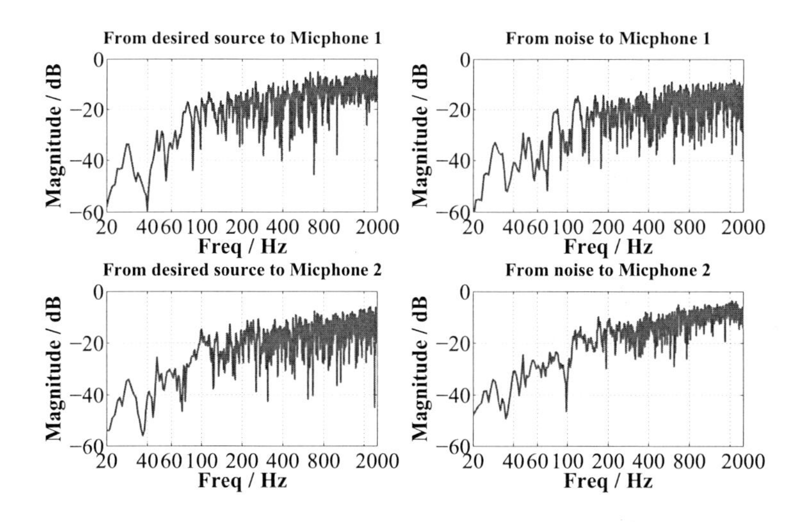

Figure 4.12.: Simulated transfer function from the source of the excitation signal and the noise source to the microphones. Two microphones are considered.

The one general separation results are presented in Fig. 4.13, and the length of the MLS sequence is 255. ICA can separate the noise in general. Since simulation is performed frequency by frequency, the following figures will present only the frequency range from 200 to 400 Hz and not lose the general properties of this statistical method.

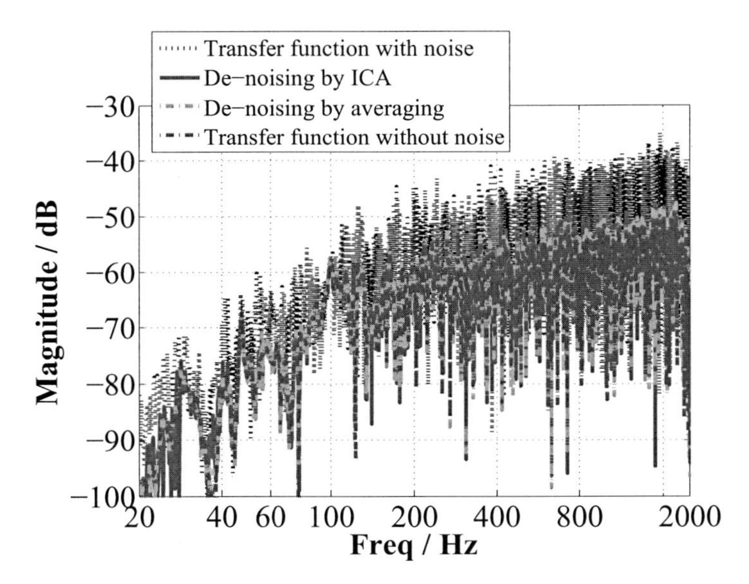

Figure 4.13.: Separation results estimated by ICA, the length of the MLS sequence is 255

Although this statistical signal separation method can be applied to separate the desired signal and the noise in general, the simulation results show that this method is unreliable and no better than averaging.

Firstly, the statistical contrast function requires a large number of samples. The samples used to compute the contrast function are the length of the MLS sequences. If insufficient samples are used in the calculation, estimation accuracy is not guaranteed. Fig. 4.14 to Fig. 4.19 show the estimation results on using ICA. As the length of the MLS sequence is only seven, the estimation is not correct. As the length of the MLS sequence increases, the errors of the ICA estimation are generally reduced.

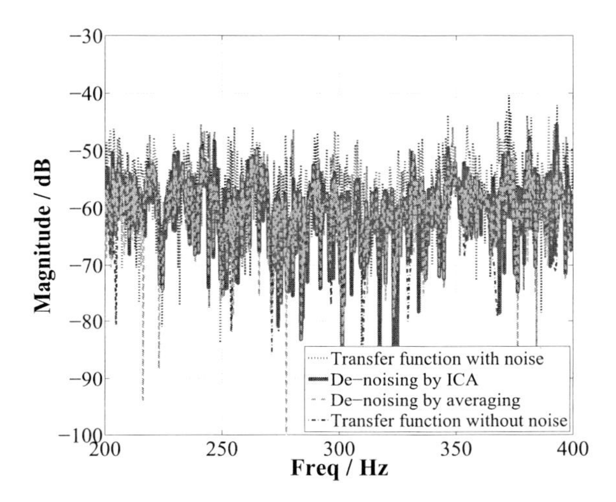

Figure 4.14.: Separation results estimated by ICA, the length of the MLS sequence is 7

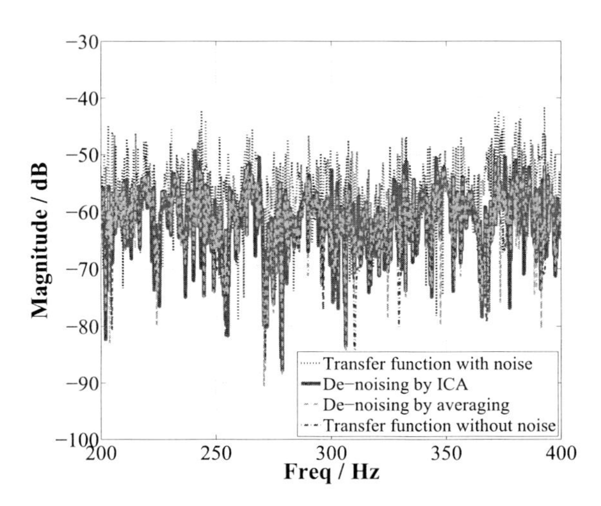

Figure 4.15.: Separation results estimated by ICA, the length of the MLS sequence is 15

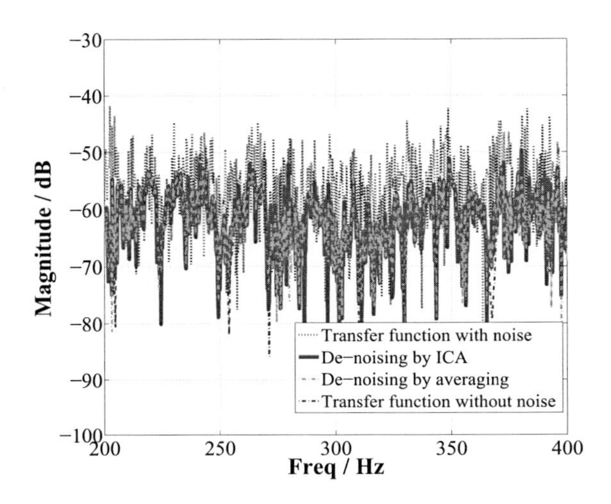

Figure 4.16.: Separation results estimated by ICA, the length of the MLS sequence is 31

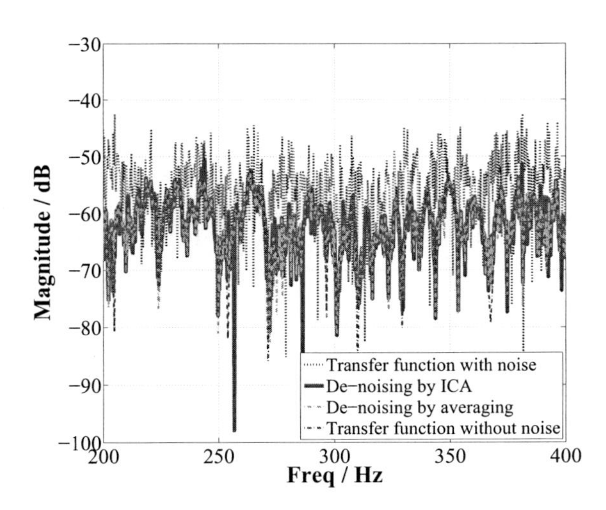

Figure 4.17.: Separation results estimated by ICA, the length of the MLS sequence is 63

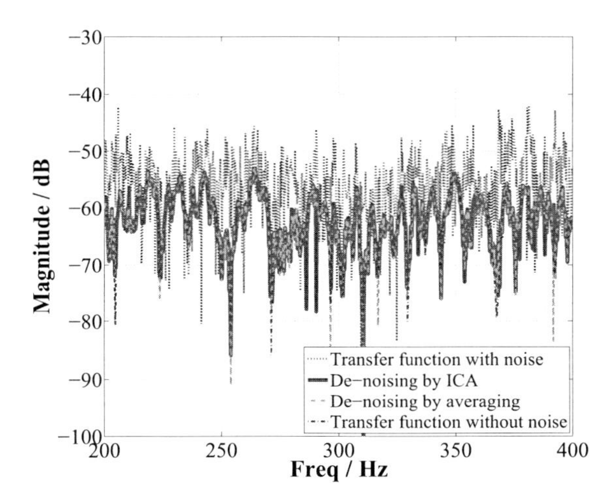

Figure 4.18.: Separation results estimated by ICA, the length of the MLS sequence is 127

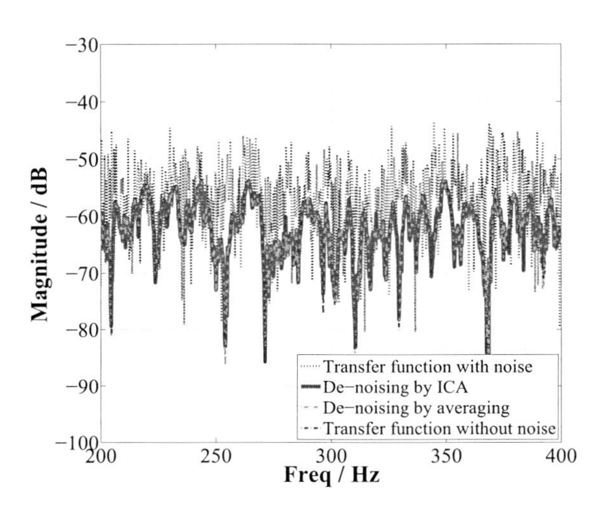

Figure 4.19.: Separation results estimated by ICA, the length of the MLS sequence is 255

If the MLS is too short—e.g. seven—the signals have insufficient samples and the signal separation is incorrect, but averaging reliably improves the SNR by approximately 9 dB. When the number of samples increases, estimation errors decrease. When the MLS sequence has 255 samples, the SNR improvement of averaging is 24 dB. The transfer function obtained by averaging is very close to the real transfer function without noise. Fig. 4.19 shows that the transfer function obtained by ICA is identical to the average approach at most of the frequencies. However, at certain frequencies—e.g. 280Hz and 380Hz—ICA shows worse results than averaging does. This is because the contrast function is theoretically satisfied when the signal is infinitely long. 255 samples are not infinite. The statistical variation in the simulated white noise of finite length leads to the error of the contrast function, and the resulting estimation of the unmixing matrix is incorrect.

Secondly, if the estimated element in the mixing matrix is small compared with other elements, the estimation errors tend to be larger. The simulation is repeated 10 times, and the mixing matrices that correspond to the transfer function of Fig. 4.12 are kept constant. The purpose is to estimate the transfer function from the source to Microphone 1. A new white noise sequence is generated for each simulation. At a few specific frequencies—such as 255 Hz, 290 Hz, 310 Hz and so on—the errors tend to be larger, sometimes, as large as 10 dB. The errors are not a constant for each simulation as shown in Fig. 4.20. Those frequencies where estimation errors tend to be larger have one property in common—the magnitude at those frequencies are small.

This is because ICA computes the statistical properties between the signals of two measurement channels. If in one measurement channel one source is very weak and the other source dominates the channel's energy, the stronger source will dominate the statistical property of this channel, and estimation errors for the weaker source will be larger. One example is presented below. Regarding the mixing equations(Eq. 4.8), the mixing matrix is arbitrarily chosen as Fig. 4.10

$$\begin{bmatrix} x_1 \\ x_2 \end{bmatrix} \begin{bmatrix} h_{11} & 1-i \\ 3-2i & 3+5i \end{bmatrix} \begin{bmatrix} \mathsf{MLS} \\ \mathsf{z}_2 \end{bmatrix} \tag{4.10}$$

h_{11} is chosen from 0.001 to 1,000. For each choice of h_{11}, the simulation is repeated 20 times, and a new white noise sequence z_2 is generated for each simulation. The length of the MLS sequence is 255. The estimation error is defined as

$$\mathrm{Err} = 20\log_{10}\left|\frac{h_{11\mathrm{estimation}} - h_{11}}{h_{11}}\right| \tag{4.11}$$

The estimation errors are presented in Fig. 4.21. The estimation errors tend to be larger when the h_{11} is small, and the errors tend to be smaller when h_{11} is larger.

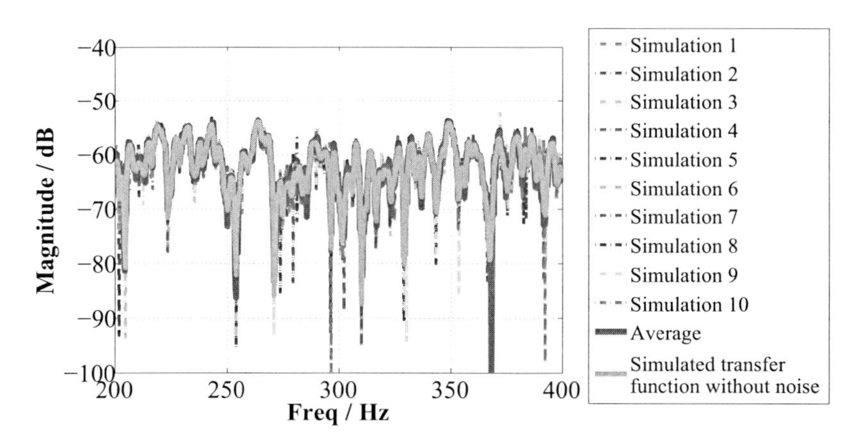

Figure 4.20.: Comparison of the averaging and the ICA. The ICA tends to have larger estimation errors at the frequencies where the magnitude of the transfer function is small.

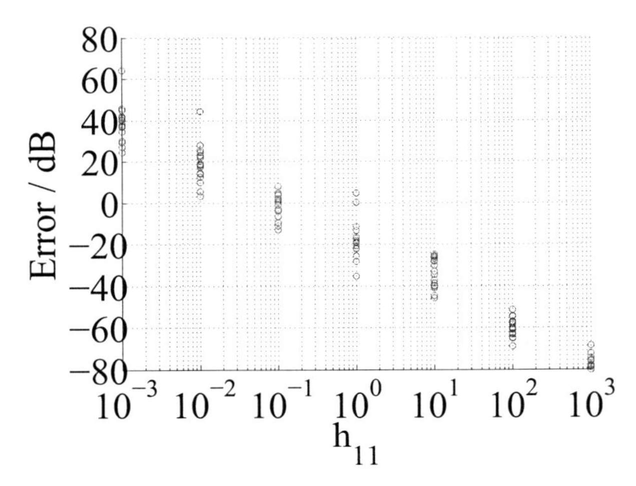

Figure 4.21.: The estimation error of h_{11} in Eq.4.10. The simulation is repeated 20 times for each arbitrary h_{11}, and a new white noise sequence is generated per repetition. The total samples are 255. The estimation errors tend to be larger when h_{11} is small.

4.5. Conclusion

This chapter aims to separate the source signal and the noise signal on impulse response measurement. The BSS-based statistical method can only deal with the simple system where only a few parameters are estimated, and cannot separate the noise in the reverberant system by a single measurement because too many parameters need to be estimated. When multiple sweeps are used to increase the number of available samples, the desired signal can be separated in principle.

Compared with conventional averaging, the statistical source separation method has to estimate the noise sequences first, and then the desired parameters. Many difficulties can lead to instability of the estimation—such as the insufficient samples, small magnitude of the desired transfer function, and the condition number of the mixing matrix that is discussed in Chapter 3. If the estimation of the noise sequence is inaccurate, the estimation error is accumulated into the final estimation of the impulse response or transfer function. When the number of noise sources increases, the statistical source separation method becomes practically prohibitive. The averaging approach estimates directly and only the desired transfer function with the least mean square errors, and is far more efficient and reliable. Hence, the idea of using the statistical source separation method for impulse response measurement should be discarded.

5

Time-variant system: wind variance

Chapter 4 shows that the averaging approach is more efficient and reliable than the source separation method. However, averaging is still not an all-conquering method. Implementation of long-time averaging runs the risk of time variance. Regarding the sound barrier measurement outdoors that was discussed in Chapter 3, the test signals are normally excited 16 times for synchronously averaging [Garai, 2011], and the overall measurement time could be five minutes. Wind variances typically occur on a time scale of several seconds. Therefore, the signals must be averaged in wind-variant systems. Thus, the problem involves establishing a robust method to compensate for the time-variance effect. Since the wind changes the phase of the measured signal, the basic approach is trying to compensate for the phase shift before averaging. The phase shift can be derived from the wave equation.

5.1. Phase shift in uniform flow

The measured signal shows a phase shift in the presence of wind. The phase shift can be derived by examining the wave equation. Considering a simple case, if there is a uniform flow along the x direction, the wave equation changes to the convective wave equation [Mechel, 2008].

$$\Delta\Psi - \frac{1}{c_0^2}\left(\frac{\partial}{\partial t} + V\frac{\partial}{\partial x}\right)^2 \Psi = 0 \tag{5.1}$$

where Ψ is the velocity potential. The wind is along the x direction (Fig. 5.1). The wave equations for the y and z directions are the general harmonic wave equation.

$$\begin{aligned} \frac{\partial^2 \Psi}{\partial y^2} - \frac{1}{c_0^2}\frac{\partial^2 \Psi}{\partial t^2} = 0 \\ \frac{\partial^2 \Psi}{\partial z^2} - \frac{1}{c_0^2}\frac{\partial^2 \Psi}{\partial t^2} = 0 \end{aligned} \tag{5.2}$$

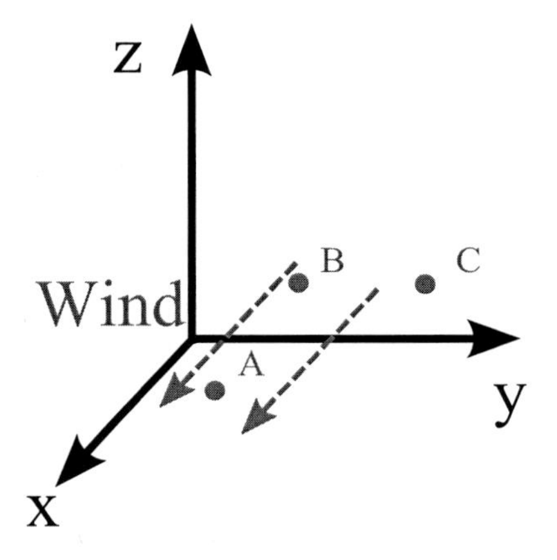

Figure 5.1.: The geometry for the wave equation. The wind is along the x direction

The point source solution in uniform wind is Eq. 5.3.

$$\Psi(r,t) = \gamma \frac{e^{i\omega\left(t + \frac{V\gamma^2 x}{c^2} - \gamma\frac{\sqrt{\gamma^2 x^2 + y^2 + z^2}}{c}\right)}}{4\pi\sqrt{\gamma^2 x^2 + y^2 + z^2}} \tag{5.3}$$

where γ is the Lorentz factor $\gamma = \frac{1}{\sqrt{1 - \frac{V^2}{c^2}}}$.

The detailed derivation is shown in Appendix. A.

If no wind ($V = 0$) is present, the Eq. 5.3 is simplified to the normal point source solution without wind.

$$\Psi(r,t) = \frac{e^{i\omega\left(t-\frac{\sqrt{x^2+y^2+z^2}}{c}\right)}}{4\pi\sqrt{x^2+y^2+z^2}} \tag{5.4}$$

Comparing the two solutions, Eq. 5.3 and Eq. 5.4 show that the phase shifts change under wind conditions, but not the frequency. If the impulse response is measured in the presence of wind, the wind variance could be compensated for by shifting the phase back to a constant group delay.

In the example of the tailwind scenario, the source is located at Point **B** in Fig. 5.1 and the receiver is located at Point **A**. In accordance with Eq. 5.3, the phase shift for tailwind is simplified as

$$\Psi(r,t) = \frac{e^{i\omega\left(t-\frac{x}{c+V}\right)}}{4\pi x} \tag{5.5}$$

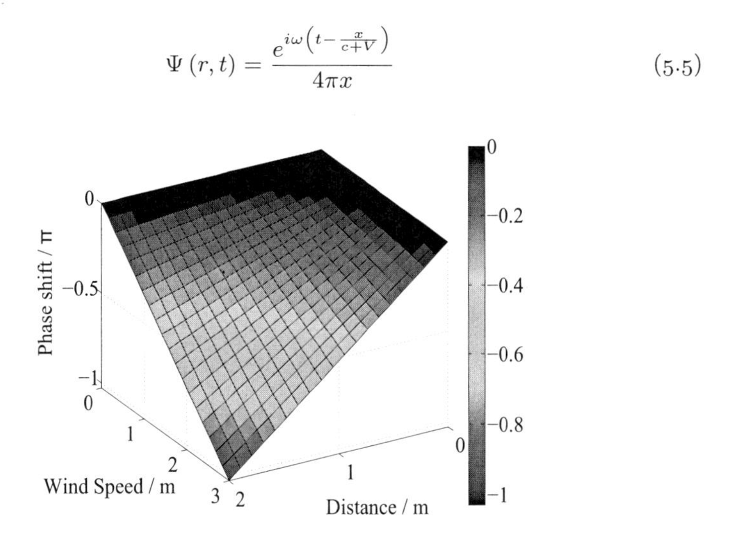

Figure 5.2.: Phase shift for tailwind at 10 kHz with respect to the wind speed and the source-receiver distance

The phase shift with respect to the frequencies and the wind speed is depicted in Fig. 5.3. When the tailwind speed is 4 m/s and the source-receiver distance is 2 m/s, the phase shift might be π over 8,000 Hz. If the average approach is implemented with such a large phase shift, the results are incorrect.

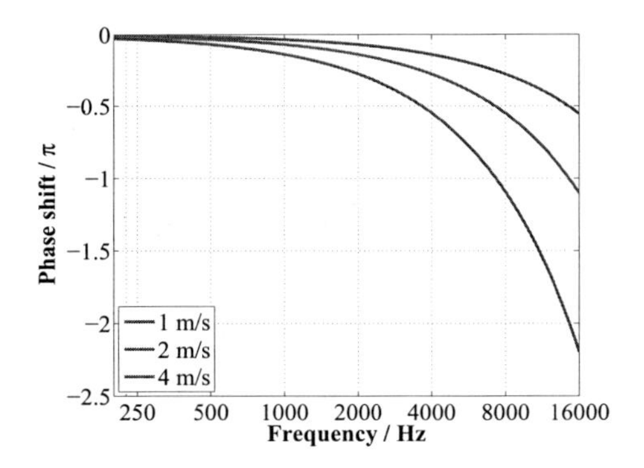

Figure 5.3.: The phase shift with respect to the frequencies caused by the tailwind. The source-receiver distance is 2 m.

If the averaging is arbitrarily implemented, an empirical acceptable phase shift between two measurements should be less than 0.07 π. In this regard, the errors after averaging will be less than 0.1 dB. To satisfy this requirement, if the wind is 4 m/s, the frequency range must be below 1,000 Hz as shown in Fig. 5.4. The magnitude change after averaging in Fig. 5.4 is calculated by $20\log_{10}\frac{1-e^{i\phi_{\text{phase shift}}}}{2}$.

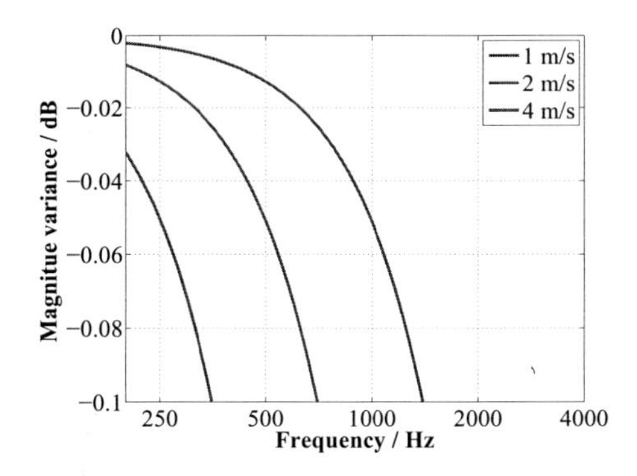

Figure 5.4.: The resulting magnitude change after averaging. The magnitude change is acceptable for averaging only at very low frequencies.

If the wind is a crosswind ($x = 0, z = 0$ in Eq.5.3), the phase shift is relatively small, proportional only to the factor $\gamma = \frac{1}{\sqrt{1-\frac{V^2}{c^2}}}$. The resulting magnitude variance is also relatively small and is plotted in Fig. 5.5.

$$\Psi(r,t) = \gamma \frac{e^{i\omega\left(t-\gamma\frac{y}{c}\right)}}{4\pi y} \tag{5.6}$$

The phase-shift variance leads to a change in magnitude before the averaging. This variance is relatively small, however, and can be neglected as shown in Fig. 5.5. The magnitude variance after averaging is calculated by $20\log_{10}\frac{1-e^{i\phi_{\text{phase shift}}}}{2}$.

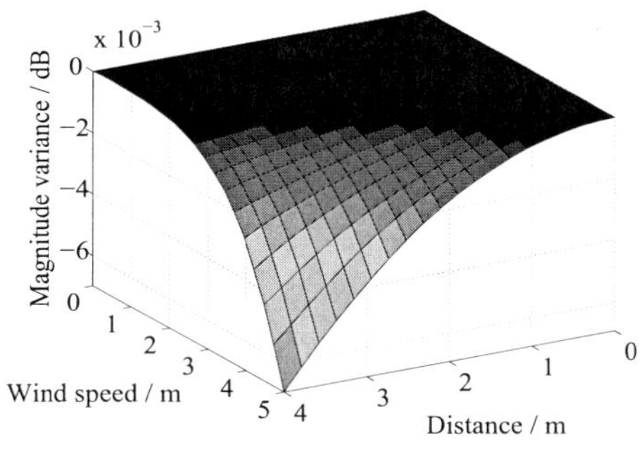

Figure 5.5.: The magnitude variance at 10 kHz caused by the phase shift. The phase shift results from the crosswind with respect to the wind speed and the source-receiver distance. The magnitude variance is only 0.006 dB when the wind speed is 5 m/s and the source-receiver distance is 4 m.

If the wind speed is known, it can be directly converted into a phase shift. Wind sensors deliver data at certain observation points, but not along the entire propagation path. In practical cases outdoors, the overall effect of the wind speed is unknown during the measurement. It has to be estimated from the measured signal.

Wind is a rapidly changing physical quantity. In the typical sound barrier measurement, the excitation signal can be tens of seconds long and at least 16 averages should be computed to maintain good background noise immunity [Garai, 2011]. In this regard, the wind varies within one excitation period.

Therefore, intra-period time variances are mainly to be studied. It is now assumed that the long excitation signal is truncated into a number of short segments, of about 1–2 ms per segment. Although the wind speed varies quickly, it can still be regarded as approximately constant within these short segments. The short segments are denoted as (tn; tn+1) for the n-th segment; therefore, the measured signal changes as given in Eq. 5.7.

$$\begin{aligned} &\text{Measurement 1:} \quad s_1(t) && t \in (t_n, t_{n+1}) \\ &\text{Measurement 2:} \quad s_2(t) = s_1(t-\tau) && t \in (t_n, t_{n+1}) \end{aligned} \tag{5.7}$$

where (t_n, t_{n+1}) corresponds to the n-th segment. The phase shift can be easily estimated by maximizing the cross correlation function.

$$Max\left\{ R(\tau_{est}) = \int_{t_n}^{t_n+1} s_1(t)s_2(t-\tau_{est})\,\mathrm{d}t \right\} \tag{5.8}$$

In this way, the intra-period phase shift in the entire measurement period can be estimated segment by segment, and the overall phase shift with respect to time $\tau(t)$ can be obtained by interpolation. In order to compensate for the phase shift smoothly, the advantage of the sweep can be used. The instant frequency of the sweep increases monotonically with time, and the time and the frequency of the sweep have one-to-one correspondence. Thus, the time-variant phase shift $\tau_{est}(t)$ can be regarded as the frequency-variant phase shift $\tau_{est}(\omega)$. Both the linear sweep and the logarithmic sweep can be used. The instant frequency of the linear sweep is proportional to the instant time, and the instant frequency of the logarithmic sweep is exponential to the instant time. Therefore, the phase difference between the two measurements can be directly compensated for in the frequency domain, as illustrated in Fig.5.6.

$$S_{2,\text{est}}(\omega) = S_2(\omega)e^{-i\omega\tau_{\text{est}}(\omega)} \tag{5.9}$$

Afterwards, synchronized time averaging can be implemented.

An experiment is performed to test the validity of the phase compensation model described above. Firstly, the single direct sound scenario is tested. The measurement setup is depicted in Fig. 5.7. The microphone, the loudspeaker and

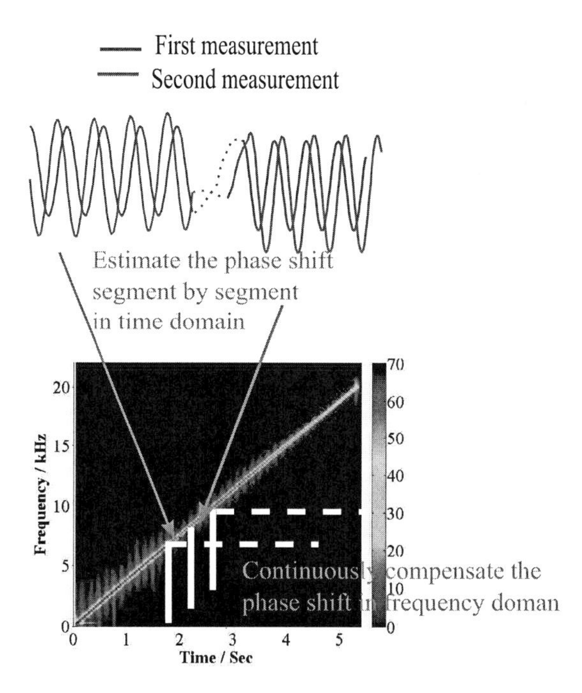

Figure 5.6.: Intra-period phase shift estimation and compensation. The continuous signals are arbitrarily truncated into discrete segments. The length of each segment used here is 50 ms. The former segment might be a negative phase shift, and the latter segments might be a positive phase shift. To avoid discontinuity after the compensation, the phase shift is interpolated and compensated for in the frequency domain.

the ventilator are positioned in a straight line. The ventilator is placed behind the loudspeaker to create a tailwind. The distance between the loudspeaker and the microphone is 2 m. This ventilator cannot generate a wind with constant speed and the speed of wind fluctuation cannot be avoided. The time-invariant case is firstly measured as a reference, and the length of the sweep is set to five seconds. The time shift and wind speed estimation results are illustrated in Fig.5.8. When synchronous averaging is performed, the apparent level loss can be observed (Fig. 5.9). As expected, the level loss effect does not occur at low frequencies because of the small phase shifts corresponding to the large wavelengths. At higher frequencies, the apparent level loss goes up to 5 dB. After phase compensation, it could be notably reduced to around 1 dB.

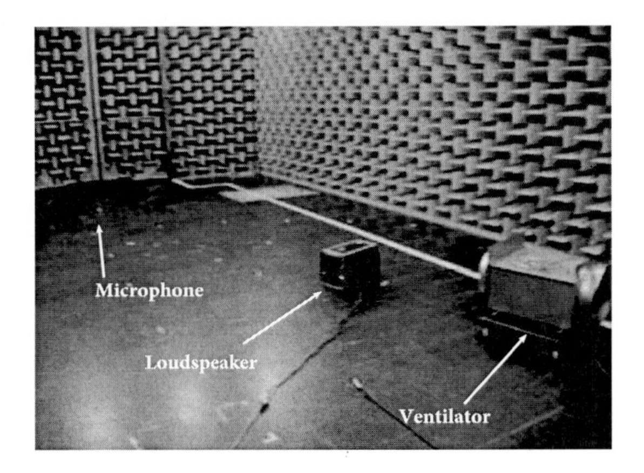

Figure 5.7.: Direct sound measurement with presence of wind

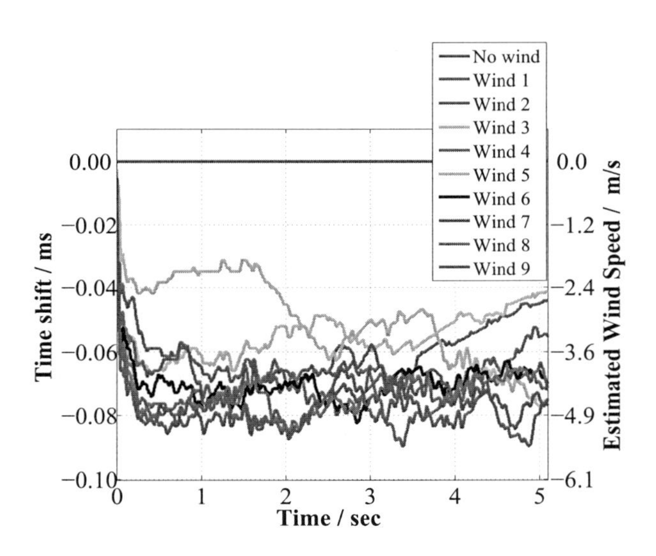

Figure 5.8.: Estimated time shift and wind speed for direct sound measurement

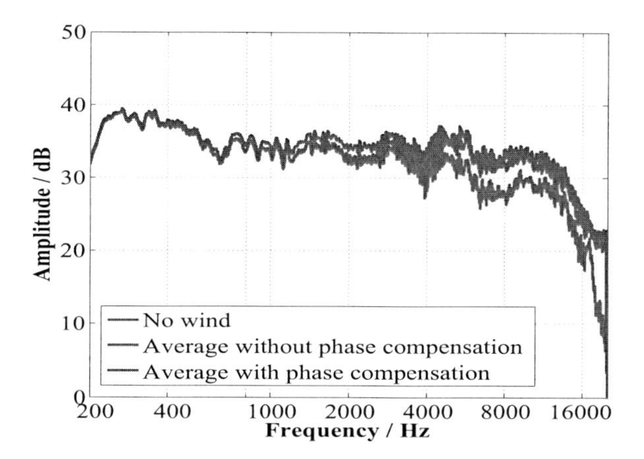

Figure 5.9.: Comparison of the average with/without phase compensation

5.2. Phase compensation for reflection coefficient

For the reflection coefficient measurement, the overall impulse response includes the direct sound and the reflected sound. The measurement for this approach is described in Chapter 3.3, and the figure of the impulse response (Fig.5.10) is plotted here once again for convenience.

The microphone is 0.75 m away from the sound barrier, and the impulse response of the reflected sound is obtained through time windowing. This measurement is performed in the anechoic chamber (Fig. 5.11), the small sound barrier model is placed close to the wall of the anechoic chamber. The wind can no longer be considered as a uniform flow. In any case, in real-life sound barrier measurements outdoors, the wind cannot be assumed to be uniform.

The transfer functions of the reflected sound measured by different winds is illustrated in Fig. 5.12. Both the phase and the magnitude change simultaneously because there are more complex effects. The magnitude varies by around 5 dB at high frequencies. The correct impulse response cannot be obtained by just compensating for the phase.

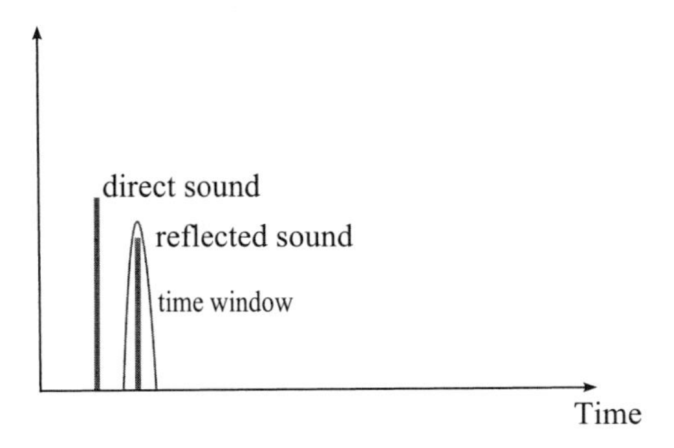

Figure 5.10.: The impulse response for sound barrier measurement. The overall impulse response includes a direct sound and a reflected sound. The reflected sound can be obtained by windowing out the impulse of direct sound if the microphone is far away from the surface under test.

Figure 5.11.: Measurement setup for sound barrier measurement with the presence of wind

5.3. Conclusion

The sound barrier must be measured under no-airflow condition. If the wind blows against the sound barrier, it is not a uniform flow. There are more complex effects, and both the magnitude and the phase are changed. No compensation is possible for this effect.

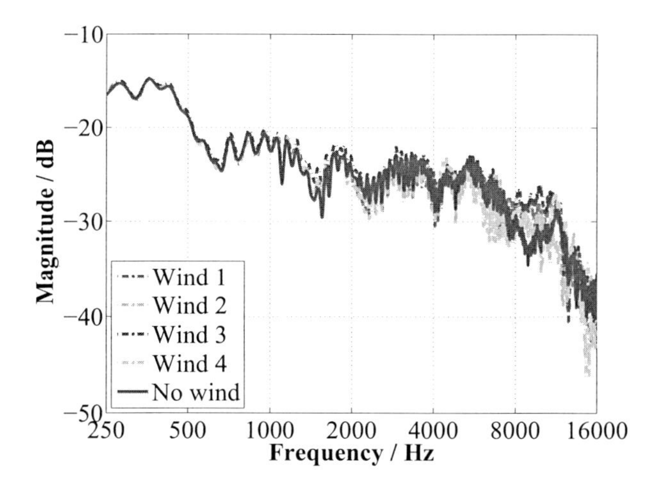

Figure 5.12.: The transfer function of the reflected sound measured by different winds. Impulse response of the reflected sound is obtained by time windowing. Since both, the magnitude and phase, change by around 5 dB at high frequencies, the correct impulse response cannot be obtained by just compensating for the phase.

6 Time-variant system: temperature variance

Another uncertainty of the impulse response measurement is the temperature variance. There is limited information, however, about the effects of temperature variances. This chapter focuses on condition monitoring of a machine structure. The temperature variance must be carefully considered during the run and subsequently heat-up of a machine. In this machine diagnosis scenario, vibration sensors and airborne sensors are used. For fault detection and for solving the inverse problem of identifying the cause of the fault, the impulse response in a machine from one component to the sensor position has to be measured. For example, machine cavities may be monitored in the production line. Different cavities of the same type must be checked for compliance with the required range of deviation from a norm. Under such conditions, temperature variances of several degrees may occur. In order to obtain accurate impulse responses, the temperature variances must be considered.

6.1. Time-stretching model

Since impulse responses are based on the wave equation, the impulse response variance caused by the temperature drift can be derived using the wave equation with boundary conditions and a temperature-dependent speed of sound. As the temperature changes, the speed of sound and the wave equation change consequently. Assuming that the temperature changes from T_0 to T_{new}, the speed of sound will change from c_0 to $\zeta \cdot c_0$, where $\zeta = \sqrt{\frac{T_{\text{new}}}{T_0}}$. The wave equation and the boundary condition are rewritten as Eq. 6.1.

$$\begin{cases} \nabla^2\Psi - \frac{1}{(\zeta\cdot c_0)^2}\frac{\partial^2\Psi}{\partial t^2} = 0 \\ c_0 = \sqrt{\frac{\gamma R T_0}{M}} \\ \rho_0\frac{\partial\Psi}{\partial t} + (\vec{n}\cdot\nabla\Psi)Z\Big|_{\varphi(\vec{r})} = 0 \end{cases} \tag{6.1}$$

Since the impulse response $h(t)$ is determined by the wave equation, and the mathematical expression of the impulse response $h(t)$ is only a function with respect to the time, an intuitive consideration could be that as the speed of sound increases, the time scale of impulse response $h(t)$ will be compressed. Performing a coordinate transform Eq. 6.2 on wave equation Eq. 6.1

$$t' = \zeta\cdot t \tag{6.2}$$

The new wave equation and the corresponding boundary condition become Eq. 6.3.

$$\begin{cases} \nabla^2\Psi - \frac{1}{c_0^2}\frac{\partial^2\Psi}{\partial t'^2} = 0 \\ c_0 = \sqrt{\frac{\gamma R T_0}{M}} \\ \rho_0\frac{\partial\Psi}{\partial t'}\cdot\zeta + (\vec{n}\cdot\nabla\Psi)Z\Big|_{\varphi(\vec{r})} = 0 \end{cases} \tag{6.3}$$

The wave equation in Eq. 6.3 is identical to Eq. 2.6. If the temperature changes, the impulse response at the new temperature should change to

$$h_{\text{new}}(t) \sim h(\zeta\cdot t) \tag{6.4}$$

Compared with Eq. 2.7, the boundary condition in Eq. 6.3 changes slightly as well. It seems that the time-stretching model presented above, Eq. 6.4, is not exactly correct. But it can be also proven that, if the impedance of the solid boundary is significantly larger than air impedance ($Z \gg \rho_0 c$) and the temperature variance is very small (e.g. only several degree Celsius), the influence of the boundary condition variance is relatively small and can be safely ignored.

First, the dependence that the direct sound component changes with the temperature is illustrated. Direct sound is measured as shown in Fig. 6.1.

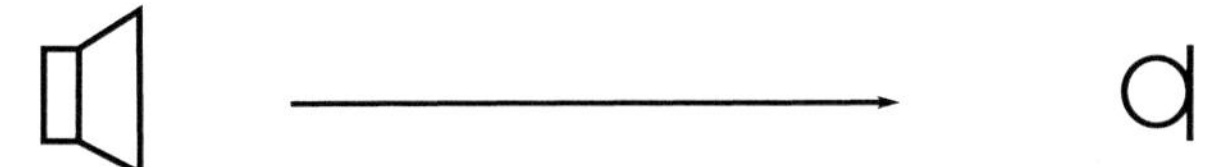

Figure 6.1.: Direct Sound

The impulse response of this system is a Dirac function.

$$h(t) = \delta(t - \frac{d}{c}) \tag{6.5}$$

As the speed of sound changes from c to $\zeta \cdot c$, the impulse response changes to

$$h_{\text{new}}(t) = \delta(t - \frac{d}{\zeta \cdot c}) = \zeta h(\zeta \cdot t) \tag{6.6}$$

The factor ζ is present in front of $h(\zeta \cdot t)$ simply because the energy density in the air increases as the temperature increases. Ignoring this energy conservation factor ζ in front of $h(\zeta \cdot t)$, the time-stretching model holds in this direct sound system.

6.1.1. Simple reflection

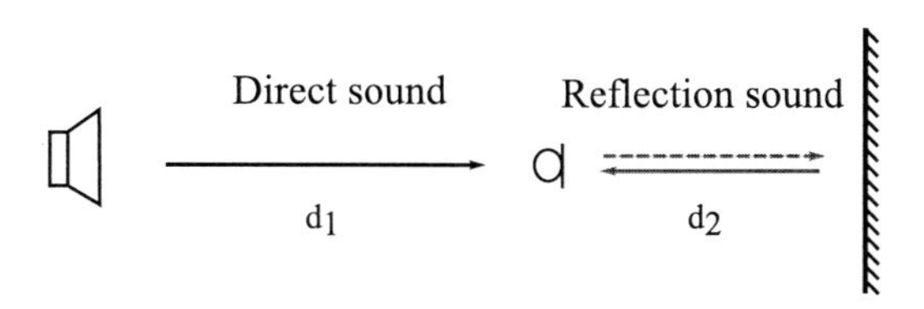

Figure 6.2.: Simple reflected sound

If only a flat wall is present, impulse response contains a direct sound and a reflection sound.

$$h(t) = \delta(t - \frac{d_1}{c}) + R\delta(t - \frac{d_1 + 2d_2}{c}) \tag{6.7}$$

where $R = \frac{Z - \rho_0 c}{Z + \rho_0 c}$. If the speed of sound changes from c to $\zeta \cdot c$, the new impulse response is

$$h_{\text{new}}(t) = \delta(t - \frac{d_1}{\zeta \cdot c}) + R_{\text{new}} \delta(t - \frac{d_1 + 2d_2}{\zeta \cdot c}) \tag{6.8}$$

where

$$R_{\text{new}} = Z \frac{1 - \frac{\zeta \cdot \rho_0 c}{Z}}{1 + \frac{\zeta \cdot \rho_0 c}{Z}} = R - \frac{1}{2}(1 - R^2)(\zeta - 1) + O(\zeta - 1)^2 \tag{6.9}$$

It seems that the new reflection factor $R_{\text{new}} \neq R$. Then the variance of the reflection factor is notated as

$$Error = 20 \log_{10} \left| \frac{R_{\text{new}} - R}{R} \right| = 20 \log_{10} \left| \frac{-1 + R^2}{2R} (\zeta - 1) + O(\zeta - 1)^2 \right| \tag{6.10}$$

But if the impedance of the solid boundary is much larger than the air impedance $Z \gg \rho_0 c$, $R \approx 1$,and $Error \approx 0$, then the time-stretching model is still applicable. The reflection variance with respect to the temperature shift and the surface impedance is shown in Fig. 6.3. The x axis is the surface impedance. Only if the surface impedance is very close to the air impedance—i.e. 415 Pa $\cdot$ s/m—does the reflection factor change considerably. For the general solid boundary, where the impedance is only at the scale of 10^5 and 10^6 Pa $\cdot$ s/m, the reflection factor changes to a lesser extent than -60 dB. The scale of **dB** is used here for convenience. -60 dB means that the reflection factor changes only 0.1%. Table 6.1 lists the characteristic impedance of several materials [Lawrence E. Kinsler, 1982]. For example, in case of wood cork, if the temperature changes from 20° C to 25° C, the error of the reflection factor obtained from Eq. 6.10 is only 0.01%, which can be safely ignored.

Solid	Characteristic Impedance (Pa · s/m)
Steel	47.0×10^6
Glass(Pyrex)	12.9×10^6
Concrete	8.0×10^6
Wood cork	0.12×10^6
Wood Pine	1.57×10^6
Rubber(hard)	2.64×10^6
Rubber(soft)	1.0×10^6
Air (20°C)	0.415×10^3

Table 6.1.: Characteristic impedance of matter

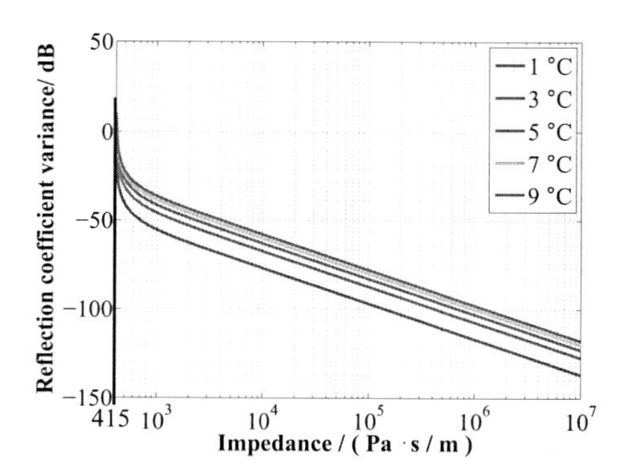

Figure 6.3.: The reflection coefficient variance with respect to wall impedance and the temperature variance. The reference temperature is assumed to be 20°C. If the temperature varies by only several degree Celsius, as long as the boundary is the normal solid material $Z > 10^5 Pa \cdot s/m$, the variance of the reflection coefficient is very small and less than -60 dB.

6.1.2. 3-dimensional scenario

The sound waves generated by a source can be expressed by the Helmholtz-Huygens integral as Eq. 6.11.

$$\begin{aligned}\Psi(\vec{r},\omega) = &\iiint_{\Omega} Q(\vec{r_0})G(\vec{r},\vec{r_0},\omega)\mathrm{d}\vec{r_0} \\ &- \iint_{S} \left[G(\vec{r},\vec{r'},\omega)\frac{\partial\Psi(\vec{r'},\omega)}{\partial n} - \Psi(\vec{r'},\omega)\frac{\partial G(\vec{r},\vec{r'},\omega)}{\partial n} \right] \mathrm{d}S\end{aligned} \tag{6.11}$$

where $\Psi(\vec{r},\omega)$ denotes the velocity potential for frequency ω, and $Q(\vec{r_0})$ is the source function. If the sound source $Q(\vec{r_0})$ is a Dirac point source, the solution for $\Psi(\vec{r},\omega)$ is directly the transfer function $H(\vec{r},\vec{r_0},\omega)$ of this system. Then the inverse Fourier transform of $\Psi(\vec{r},\omega)$ will be the impulse response $h(\vec{r},\vec{r_0},t)$ of this system. The first integration $\iiint_{\Omega} Q(\vec{r_0})G(\vec{r},\vec{r_0},\omega)\mathrm{d}\vec{r_0}$ stands for the direct sound from the source. The second integration is the contribution from the reflection caused by the boundary, where $\vec{r'}$ stands for the boundary's coordinate. $G(\vec{r},\vec{r'},\omega)$ and $G(\vec{r},\vec{r_0},\omega)$ are the Green's functions for a point source in free space (Eq. 6.12), which is the solution of the inhomogeneous Helmholtz Equation in free space (Eq. 6.13).

$$G(\vec{r},\vec{r_0},\omega) = \frac{e^{-ikr}}{4\pi r} = \frac{e^{-ik\sqrt{(x-x_0)^2+(y-y_0)^2+(z-z_0)^2}}}{4\pi\sqrt{(x-x_0)^2+(y-y_0)^2+(z-z_0)^2}} \tag{6.12}$$

where $k = \frac{\omega}{c_0}$.

$$\nabla^2 G(\vec{r},\vec{r_0},\omega) + \frac{\omega^2}{c_0^2}G(\vec{r},\vec{r_0},\omega) = -\delta(\vec{r}-\vec{r_0}) \tag{6.13}$$

Considering the boundary condition in Eq. 6.1

$$\frac{\partial\Psi(\vec{r'},\omega)}{\partial n} = (\vec{n}\cdot\nabla\Psi) = -\rho_0\frac{\partial\Psi}{\partial t} = -\frac{\rho_0}{Z}\cdot i\omega\cdot\Psi = -\frac{Z_0}{Z}\cdot i\frac{\omega}{c_0}\Psi \tag{6.14}$$

where Z_0 is the impedance of the air , $Z_0 = \rho_0 c_0$. Then

$$\begin{aligned}\Psi(\vec{r},\omega) &= \iiint_{\Omega} Q(\vec{r_0})G(\vec{r},\vec{r_0},\omega)\mathrm{d}\vec{r_0} \\ &\quad - \iint_{S}\left[-\frac{Z_0}{Z}\cdot i\frac{\omega}{c_0}\cdot\Psi(\vec{r'},\omega)G(\vec{r},\vec{r'},\omega) - \Psi(\vec{r'},\omega)\frac{\partial G(\vec{r},\vec{r'},\omega)}{\partial n}\right]\mathrm{d}S \\ &= \iiint_{\Omega} Q(\vec{r_0})G(\vec{r},\vec{r_0},\omega)\mathrm{d}\vec{r_0} \\ &\quad + \iint_{S}\left[\frac{Z_0}{Z}\cdot i\frac{\omega}{c_0}\cdot G(\vec{r},\vec{r'},\omega) + \frac{\partial G(\vec{r},\vec{r'},\omega)}{\partial n}\right]\Psi(\vec{r'},\omega)\mathrm{d}S \end{aligned} \tag{6.15}$$

If the speed of sound changes from c_0 to a new speed $\zeta \cdot c_0$, the inhomogeneous Helmholtz equation in free space changes to

$$\nabla^2 G_{\mathrm{new}}(\vec{r},\vec{r_0},\omega) + \frac{\omega^2}{(\zeta\cdot c_0)^2}G_{\mathrm{new}}(\vec{r},\vec{r_0},\omega) = -\delta(\vec{r}-\vec{r_0}) \tag{6.16}$$

Green's function for the point source changes to

$$G_{\mathrm{new}}(\vec{r},\vec{r_0},\omega) = \frac{e^{-ik_{\mathrm{new}}r}}{4\pi r} = \frac{e^{-i\frac{\omega}{\zeta\cdot c_0}r}}{4\pi r} = G(\vec{r},\vec{r_0},\frac{\omega}{\zeta}) \tag{6.17}$$

where $k_{\mathrm{new}} = \frac{\omega}{\zeta\cdot c_0}$.

The boundary condition at the new temperature can still be written as

$$\frac{\partial\Psi_{\mathrm{new}}(\vec{r'},\omega)}{\partial n} = (\vec{n}\cdot\nabla\Psi_{\mathrm{new}}) = -\rho_0\frac{\partial\Psi_{\mathrm{new}}}{\partial t} = -\frac{\rho_0}{Z}\cdot i\omega\cdot\Psi_{\mathrm{new}} = -\frac{Z_0}{Z}\cdot i\frac{\omega}{c}\Psi_{\mathrm{new}} \tag{6.18}$$

The Helmholtz integral equation Eq. 6.15 at the new temperature changes to

$$\begin{aligned}\Psi_{\mathrm{new}}(\vec{r},\omega) &= \iiint\limits_{\Omega} Q(\vec{r_0})G_{\mathrm{new}}(\vec{r},\vec{r_0},\omega)\mathrm{d}\vec{r_0} \\ &+ \iint\limits_{S}\left[\frac{Z_0}{Z}\cdot i\frac{\omega}{c_0}\cdot G_{\mathrm{new}}(\vec{r},\vec{r'},\omega)+\frac{\partial G_{\mathrm{new}}(\vec{r},\vec{r'},\omega)}{\partial n}\right]\Psi_{\mathrm{new}}(\vec{r'},\omega)\mathrm{d}S \\ &= \iiint\limits_{\Omega} Q(\vec{r_0})G(\vec{r},\vec{r_0},\frac{\omega}{\zeta})\mathrm{d}\vec{r_0} \\ &+ \iint\limits_{S}\left[\frac{Z_0}{Z}\cdot i\frac{\omega}{c_0}\cdot G(\vec{r},\vec{r'},\frac{\omega}{\zeta})+\frac{\partial G(\vec{r},\vec{r'},\frac{\omega}{\zeta})}{\partial n}\right]\Psi_{\mathrm{new}}(\vec{r'},\omega)\mathrm{d}S\end{aligned} \tag{6.19}$$

Since the frequency variable of Green's function at the new temperature changes with a factor $\frac{\omega}{\zeta}$, the sound field at new temperature can be written as a trial solution as Eq. 6.20.

$$\Psi_{\mathrm{new}}(\vec{r},\omega) = \Psi(\vec{r},\frac{\omega}{\zeta}) + \delta\Psi(\vec{r},\omega) \tag{6.20}$$

It will be proven from Eq. 6.21 to Eq. 6.26 that once the impedance of the boundary is much larger than the air impedance and the temperature variance is small, the error function $\delta\Psi(\vec{r},\omega)$ is relatively small; hence, it can be safely ignored.

$\Psi_{\mathrm{new}}(\vec{r},\omega)$ is inserted in Eq. 6.19 with Eq. 6.20

$$\begin{aligned}
&\Psi(\vec{r},\frac{\omega}{\zeta})+\delta\Psi(\vec{r},\omega)\\
&\quad=\iiint\limits_{\Omega} Q(\vec{r_0})G(\vec{r},\vec{r_0},\frac{\omega}{\zeta})\mathrm{d}\vec{r_0}\\
&\qquad+\iint\limits_{S}\left[\frac{Z_0}{Z}\cdot i\frac{\omega}{c_0}\cdot G(\vec{r},\vec{r'},\frac{\omega}{\zeta})+\frac{\partial G(\vec{r},\vec{r'},\frac{\omega}{\zeta})}{\partial n}\right]\left[\Psi(\vec{r'},\frac{\omega}{\zeta})+\delta\Psi(\vec{r'},\omega)\right]\mathrm{d}S\\
&\quad=\iiint\limits_{\Omega} Q(\vec{r_0})G(\vec{r},\vec{r_0},\frac{\omega}{\zeta})\mathrm{d}\vec{r_0}\\
&\qquad+\iint\limits_{S}\left[\frac{Z_0}{Z}\cdot i\frac{\omega}{c_0}\cdot G(\vec{r},\vec{r'},\frac{\omega}{\zeta})+\frac{\partial G(\vec{r},\vec{r'},\frac{\omega}{\zeta})}{\partial n}\right]\Psi(\vec{r'},\frac{\omega}{\zeta})\mathrm{d}S\\
&\qquad+\iint\limits_{S}\left[\frac{Z_0}{Z}\cdot i\frac{\omega}{c_0}\cdot G(\vec{r},\vec{r'},\frac{\omega}{\zeta})+\frac{\partial G(\vec{r},\vec{r'},\frac{\omega}{\zeta})}{\partial n}\right]\delta\Psi(\vec{r'},\omega)\mathrm{d}S
\end{aligned}\tag{6.21}$$

because $\Psi(\vec{r},\omega)$ is determined by Eq. 6.15.

Therefore,

$$\begin{aligned}
\Psi(\vec{r},\frac{\omega}{\zeta})=&\iiint\limits_{\Omega} Q(\vec{r_0})G(\vec{r},\vec{r_0},\frac{\omega}{\zeta})\mathrm{d}\vec{r_0}\\
&+\iint\limits_{S}\left[\frac{Z_0}{Z}\cdot\frac{i\omega}{\zeta\cdot c_0}\cdot G(\vec{r},\vec{r'},\frac{\omega}{\zeta})+\frac{\partial G(\vec{r},\vec{r'},\frac{\omega}{\zeta})}{\partial n}\right]\Psi(\vec{r'},\frac{\omega}{\zeta})\mathrm{d}S
\end{aligned}\tag{6.22}$$

The difference in the velocity potential is given by subtracting Eq. 6.22 from Eq. 6.21.

$$\begin{aligned}\delta\Psi(\vec{r},\omega) &= \iint\limits_S \left[\frac{Z_0}{Z}\cdot i\frac{\omega}{c_0}\cdot(1-\frac{1}{\zeta})G(\vec{r},\vec{r'},\frac{\omega}{\zeta})\right]\Psi(\vec{r'},\frac{\omega}{\zeta})\mathrm{d}S \\ &+ \iint\limits_S \left[\frac{Z_0}{Z}\cdot i\frac{\omega}{c_0}\cdot G(\vec{r},\vec{r'},\frac{\omega}{\zeta}) + \frac{\partial G(\vec{r},\vec{r'},\frac{\omega}{\zeta})}{\partial n}\right]\delta\Psi(\vec{r'},\omega)\mathrm{d}S \\ &= \iint\limits_S \left[\frac{Z_0}{Z}(\zeta-1)\cdot i\frac{\omega}{\zeta\cdot c_0}\cdot G(\vec{r},\vec{r'},\frac{\omega}{\zeta})\right]\Psi(\vec{r'},\frac{\omega}{\zeta})\mathrm{d}S \\ &+ \iint\limits_S \left[\frac{Z_0}{Z}\cdot i\frac{\omega}{c}\cdot G(\vec{r},\vec{r'},\frac{\omega}{\zeta}) + \frac{\partial G(\vec{r},\vec{r'},\frac{\omega}{\zeta})}{\partial n}\right]\delta\Psi(\vec{r'},\omega)\mathrm{d}S\end{aligned} \tag{6.23}$$

From Eq. 6.23, the sound field $\delta\Psi(\vec{r},\omega)$ is equivalent to the sound field generated by an extra source δQ as Eq. 6.24.

$$\iiint\limits_\Omega \delta Q(\vec{r}_0)G(\vec{r},\vec{r}_0,\frac{\omega}{\zeta})\mathrm{d}\vec{r}_0 = \iint\limits_S \left[\frac{Z_0}{Z}(\zeta-1)\cdot i\frac{\omega}{\zeta\cdot c_0}\cdot G(\vec{r},\vec{r'},\frac{\omega}{\zeta})\right]\Psi(\vec{r'},\frac{\omega}{\zeta})\mathrm{d}S \tag{6.24}$$

Since the order of magnitude of $\frac{\partial G(\vec{r},\vec{r'},\frac{\omega}{\zeta})}{\partial n}$ is $\sim -(i\frac{\omega}{\zeta\cdot c}+\frac{1}{r})G(\vec{r},\vec{r'},\frac{\omega}{\zeta})$, the proportion $\frac{\delta\Psi(\vec{r},\omega)}{\Psi(\vec{r},\frac{\omega}{\zeta})}$ is roughly the order of magnitude of $\frac{Z_0}{Z}\cdot(\zeta-1)$. Under the condition that impedance of the boundary is much larger than the air impedance and the temperature variance is also small as in Eq. 6.25, the error function $\delta\Psi(\vec{r},\omega)$ is small when compared to the $\Psi(\vec{r},\frac{\omega}{\zeta})$.

$$\frac{Z_0}{Z}\cdot(\zeta-1)\ll 1 \tag{6.25}$$

Now, the sound field at the new temperature can be approximately written as

$$\Psi_{\mathrm{new}}(\vec{r},\omega)\approx\Psi(\vec{r},\frac{\omega}{\zeta}) \tag{6.26}$$

Hence, the transfer function at the new temperature is $H(\vec{r},\vec{r}_0,\frac{\omega}{\zeta})$. Transforming $H(\vec{r},\vec{r}_0,\frac{\omega}{\zeta})$ to the time domain results in $\zeta\cdot h(\zeta\cdot t)$. The factor ζ is present in front of $h(\zeta \cdot t)$ only because of the energy conversation of the Fourier transform. These are listed in Table. 6.2.

Absolute Temperature	T_0	T_{new}
Sound Speed	c_0	$\zeta \cdot c_0 \quad \left(\zeta = \sqrt{\frac{T_{\text{new}}}{T_0}}\right)$
Impulse Response	$h(t)$	$\zeta \cdot h(\zeta \cdot t)$
Transfer Function	$H(\omega)$	$H(\frac{\omega}{\zeta})$

Table 6.2.: Time-stretching model

6.2. Intra-period time variance

The time-stretching model derived above is for impulse response at an instantaneous time. If a very long excitation signal is generated and the temperature gradually changes during the measurement, the aforementioned time-stretching model cannot be implemented directly. This is because ζ is not a constant but a function of time, $\zeta(t)$, itself. In order to solve this intra-period time variance problem, the wave equation is considered again.

The impulse response $h(t)$ is the solution of Ψ.

When a long excitation signal is generated, the time-variant wave equation is

$$\begin{cases} \nabla^2\Psi - \frac{1}{\zeta(t)^2 c_0^2}\frac{\partial^2\Psi}{\partial t^2} = \delta(\vec{r} - \vec{r}_0)s(t) \\ c_0 = \sqrt{\frac{\gamma R T_0}{M}} \\ \zeta(t) = \sqrt{\frac{T(t)}{T_0}} \end{cases} \tag{6.27}$$

where $\zeta(t)$ indicates the speed of sound changes over time.

Then the solution of this time-variant wave equation can be approximately written as

$$\Psi(\vec{r}, t) = \int_{-\infty}^{\infty} H(\vec{r}, \vec{r}_0, \omega)A(\omega)e^{i\omega \int_0^t \zeta(t)\,\mathrm{d}t}\mathrm{d}\omega \tag{6.28}$$

where $H(\vec{r}, \vec{r}_0, \omega)$ is the transfer function of this system for $\zeta = 1$, when the temperature is T_0. This solution is the same as for the Helmholtz equation, Eq. 6.29

$$\begin{cases} \nabla^2 H(\vec{r}, \vec{r}_0, \omega) + \dfrac{\omega^2}{c_0^2} H(\vec{r}, \vec{r}_0, \omega) = \delta(\vec{r} - \vec{r}_0) \\ H(\vec{r}, \vec{r}_0, \omega) \quad \text{fits the boundary condition} \end{cases} \tag{6.29}$$

because

$$\begin{aligned} \frac{1}{\zeta(t)^2 c_0^2} \frac{\partial^2}{\partial t^2} \left(e^{i\omega \int_0^t \zeta(t)\,\mathrm{d}t} \right) &= \frac{1}{\zeta(t)^2 c_0^2} \left[-\omega^2 \zeta(t)^2 + i\omega \frac{\partial \zeta(t)}{\partial t} \right] e^{i\omega \int_0^t \zeta(t)\,\mathrm{d}t} \\ &= \frac{-\omega^2}{c_0^2} \left[1 - i \frac{1}{\omega \cdot \zeta(t)^2} \frac{\partial \zeta(t)}{\partial t} \right] e^{i\omega \int_0^t \zeta(t)\,\mathrm{d}t} \end{aligned} \tag{6.30}$$

Usually, the temperature variance $\frac{\partial \zeta}{\partial t}$ is very small. For example, at temperature 20°C, the temperature variance is 2°C/minute, and the frequency is $f = 20$ Hz $\frac{1}{\omega \cdot \zeta(t)^2} \frac{\partial \zeta(t)}{\partial t} \approx 5 \times 10^{-7}$ and $\left[1 - i \frac{1}{\omega \cdot \zeta(t)^2} \frac{\partial \zeta(t)}{\partial t} \right] = 1 - 0.0000005i \approx 1$; hence, the term $\frac{1}{\omega \cdot \zeta(t)^2} \frac{\partial \zeta(t)}{\partial t}$ can be safely neglected.

Therefore,

$$\frac{1}{\zeta(t)^2 c_0^2} \frac{\partial^2}{\partial t^2} \left(e^{i\omega \int_0^t \zeta(t)\,\mathrm{d}t} \right) \approx -\frac{\omega^2}{c_0^2} e^{i\omega \int_0^t \zeta(t)\,\mathrm{d}t} \tag{6.31}$$

$$\frac{1}{\zeta(t)^2 c_0^2} \frac{\partial^2 \Psi(\vec{r}, t)}{\partial t^2} \approx -\frac{\omega^2}{c_0^2} \int_{-\infty}^{\infty} H(\vec{r}, \vec{r}_0, \omega) A(\omega) e^{i\omega \int_0^t \zeta(t)\,\mathrm{d}t}\,\mathrm{d}\omega \tag{6.32}$$

Replace Ψ in the time-variant wave equation Eq. 6.27 with Eq.6.28.

$$
\begin{aligned}
&\nabla^2\Psi(\vec{r},t) - \frac{1}{\zeta(t)^2 c_0^2}\frac{\partial^2\Psi(\vec{r},t)}{\partial t^2} \\
&= \int_{-\infty}^{\infty} \nabla^2 H(\vec{r},\vec{r}_0,\omega)A(\omega)e^{i\omega\int_0^t \zeta(t)\,\mathrm{d}t}\mathrm{d}\omega + \frac{\omega^2}{c_0^2}\int_{-\infty}^{\infty} H(\vec{r},\vec{r}_0,\omega)A(\omega)e^{i\omega\int_0^t \zeta(t)\,\mathrm{d}t}\,\mathrm{d}\omega \\
&= \int_{-\infty}^{\infty}\left[\nabla^2 H(\vec{r},\vec{r}_0,\omega) + \frac{\omega^2}{c^2}H(\vec{r},\vec{r}_0,\omega)\right]A(\omega)e^{i\omega\int_0^t \zeta(t)\,\mathrm{d}t}\mathrm{d}\omega \\
&= \int_{-\infty}^{\infty}\delta(\vec{r}-\vec{r}_0)A(\omega)e^{i\omega\int_0^t \zeta(t)\,\mathrm{d}t}\mathrm{d}\omega \\
&= \delta(\vec{r}-\vec{r}_0)s(t)
\end{aligned}
\tag{6.33}
$$

Then

$$\int_{-\infty}^{\infty} A(\omega)e^{i\omega\int_0^t \zeta(t)\,\mathrm{d}t}\mathrm{d}\omega = s(t) \tag{6.34}$$

Assign that

$$t' = \xi(t) = \int_0^t \zeta(t)\,\mathrm{d}t \tag{6.35}$$

$$t = \xi^{-1}(t') \tag{6.36}$$

$$\int_{-\infty}^{\infty} A(\omega)e^{i\omega t'}\mathrm{d}\omega = s\left[\xi^{-1}(t')\right] \tag{6.37}$$

Let $s'(t)$ stand for the time-warped excitation signal $s\left[\xi^{-1}(t')\right]$.

Then $A(\omega) = S'(\omega)$, where $S'(\omega)$ is the Fourier transformation of $s'(t)$.

Finally, as the excitation signal $s(t)$ is generated, the sound wave is described by Eq. 6.38.

$$\Psi(\vec{r},t) = \int_{-\infty}^{\infty} H(\vec{r},\vec{r_0},\omega)S'(\omega)e^{i\omega\cdot\xi(t)}\mathrm{d}\omega \tag{6.38}$$

Eq. 6.38 can be explained by the convolution theory. The transfer function is the Fourier transform of the impulse response as in Eq. 6.39.

$$H(\vec{r},\vec{r_0},\omega) = \int_{-\infty}^{\infty} h(\vec{r},\vec{r_0},t)e^{-i\omega t}\mathrm{d}t \tag{6.39}$$

In addition, $S'(\omega)$ is the Fourier transform of the time-warped signal $s'(t)$.

Then Eq. 6.38 can be rewritten as

$$\begin{aligned}\Psi(\vec{r},t) &= \int_{-\infty}^{\infty}\left[\int_{-\infty}^{\infty} h(\vec{r},\vec{r_0},t_1)e^{-i\omega t_1}\mathrm{d}t_1\right]\cdot\left[\int_{-\infty}^{\infty} s'(t_2)e^{-i\omega t_2}\mathrm{d}t_2\right]e^{i\omega\cdot\xi(t)}\,\mathrm{d}\omega \\ &= \int_{-\infty}^{\infty} h\left(\vec{r},\vec{r_0},\xi(t)-t_2\right)s\left[\xi^{-1}(t)\right]\,\mathrm{d}t_2\end{aligned} \tag{6.40}$$

Eq. 6.40 is exactly the measured signal $y(t)$, when an excitation signal $s(t)$ is generated. The process looks complicated, but the following two-step process explains it clearly.

Step 1 The impulse response at temperature T_0 is convolved with the time-warped excitation signal.

$$y_{\text{step1}}(t) = h(\vec{r},\vec{r_0},t) * s\left[\xi^{-1}(t)\right] \tag{6.41}$$

Step 2 The signal of the first step is warped by factor $\xi(t) = \int_0^t \zeta(t)\,\mathrm{d}t$.

$$y(t) = y_{\text{step2}}(t) = y_{\text{step1}}\left[\xi(t)\right] \tag{6.42}$$

Once the excitation signal $s(t)$ is generated, and the measured signal $y(t)$ is recorded, if the time-dependent temperature variance factor $\zeta(t)$ is known, the correct impulse response can be recovered by deconvolving $y\left[\xi^{-1}(t)\right]$ with $s\left[\xi^{-1}(t)\right]$, as shown in Eq. 6.43.

	Physical Process	h(t) calculation
No temperature variance	$y(t) = h(t) * s(t)$	$y(t) = h(t) * s(t)$
Inter-period temperature variance	$y(t) = h(\zeta \cdot t) * s(t)$	$y(\frac{t}{\zeta}) = h(t) * s(\frac{t}{\zeta})$
Intra-period temperature variance	$y(t) = \int_{-\infty}^{\infty} h\left(\xi(t) - t_2\right) s\left[\xi^{-1}(t)\right] \mathrm{d}t_2$	$y\left[\xi^{-1}(t)\right] = h(t) * s\left[\xi^{-1}(t)\right]$

Table 6.3.: The impulse response calculation. The first line denotes whether the impulse response is without temperature variance. The third line is the impulse response of the intra-period time-variance calculation. The second line represents a particular case of the intra-period time variance where ζ is assumed to be a constant.

$$y\left[\xi^{-1}(t)\right] = h(\vec{r}, \vec{r_0}, t) * s\left[\xi^{-1}(t)\right] \tag{6.43}$$

In a specific case, if the temperature remains constant within the measurement period, as ζ , $\xi(t) = \int_0^t \zeta(t)\,\mathrm{d}t = \zeta \cdot t$, the correct impulse response can be extracted by

$$y\left(\frac{t}{\zeta}\right) = h(\vec{r}, \vec{r_0}, t) * s\left(\frac{t}{\zeta}\right) \tag{6.44}$$

This is equivalent to the instantaneous time-stretching model that was described in Section 6.1.

6.3. Estimation of the time-stretching factor

In order to obtain an estimate of the time-stretching factor $\zeta(t)$, one approach is to identify the temperature directly during the measurement. However, in order to ascertain all possible causes for time variances, a more reliable approach is to estimate the time-stretching factor from the measurement data itself. Two different cases—i.e. inter-period time variance and intra-period time variance—have to treated separately.

Inter-period time variance

Considering two measurements at different temperatures, the temperature within a single measurement is constant. But for the first measurement, the temperature is T_0. During the the second measurement, the temperature changes to a new temperature T_{new} and the time-stretching factor is $\zeta = \sqrt{\frac{T_{\text{new}}}{T_0}}$. Then the two measured signals are

$$\begin{aligned} y_1(t) &= h(t) * s(t) \\ y_2(t) &= h_{\text{new}}(t) * s(t) = \zeta \cdot h(\zeta \cdot t) * s(t) \end{aligned} \tag{6.45}$$

The time-stretching factor can be estimated by least mean square estimation, as in Eq. 6.46.

$$\min\{\int \left[h(t) - \frac{1}{\zeta_{\text{est}}} h_{\text{new}}(\frac{t}{\zeta_{\text{est}}})\right]^2 \} \, \mathrm{d}t \tag{6.46}$$

For the sake of reliability, however, the cross-correlation function is used. The time-stretching factor is estimated by maximizing the cross-correlation function Eq. 6.47.

$$R_{hh}(\zeta_{\text{est}}) = \int h(t) h_{\text{new}}(\frac{t}{\zeta_{\text{est}}}) \, \mathrm{d}t \tag{6.47}$$

Since Eq. 6.46 and Eq. 6.47 are mathematically equivalent, the advantage of the cross-correlation function estimator is that it ignores the magnitude, and the estimation concentrates on the phase shift. This makes estimation for ζ more reliable.

In order to calculate the aforementioned time-stretching process $h_{\text{new}}(\frac{t}{\zeta_{est}})$, an interpolation must be implemented. Both linear interpolation and spline interpolation can be used. In order to ensure smaller interpolation errors, the signal must first be upsampled to a sufficiently high sampling rate. For empirical considerations, the -70dB error bounds for the interpolation are sufficient for calculating the $h_{\text{new}}(\frac{t}{\zeta_{est}})$.

For example, as per [De Boor, 2001; Hall and Meyer, 1976], the error bounds for spline interpolation are approximately

$$|e(t)| \backsim \frac{5}{384}||g^{(4)}(t)||_{\infty} \cdot \tau^4 \tag{6.48}$$

where $||g^{(4)}(t)||_{\infty}$ is the maximum of the fourth-order derivative of the interpolated function over the entire interpolation interval, and $\tau$is the sampling interval. Assuming that the frequency range of the impulse response is up to 20kHz and the impulse response is measured in a 44.1kHz sampling rate,

$$||g^{(4)}(t)||_{\infty} = \max\left\{\frac{\partial^4 \sin\left(2\pi \cdot 20000 \cdot t\right)}{\partial t^4}\right\} = (2\pi \cdot 20000)^4 \tag{6.49}$$

The measured signal is upsampled to 8×44.1 kHz, and the error bounds are

$$|e(t)| \backsim \frac{5}{384} \cdot (2\pi \cdot 20000)^4 \cdot \left(\frac{1}{8 \times 44100}\right)^4 = -73\text{dB} \tag{6.50}$$

This interpolation error is interpreted similarly to the quantization noise, and it is small enough to ensure accuracy for calculating the time-stretching process $h_{\text{new}}(\frac{t}{\zeta_{est}})$.

Intra-period time variance

If the excitation signal is very long, the temperature changes within the prolonged measurement period, but it does so slowly and slightly. The long excitation signal can still be divided into short segments. Within these short segments, the temperature can be considered approximately constant, and the estimation converges to the least mean square error that can be considered the 'average' temperature within this short segment. Then the overall time-dependent stretching factor $\zeta(t)$ can be estimated by moving and averaging the time-stretching factor segment by segment using the inter-period time-stretching model, as shown in Fig. 6.4.

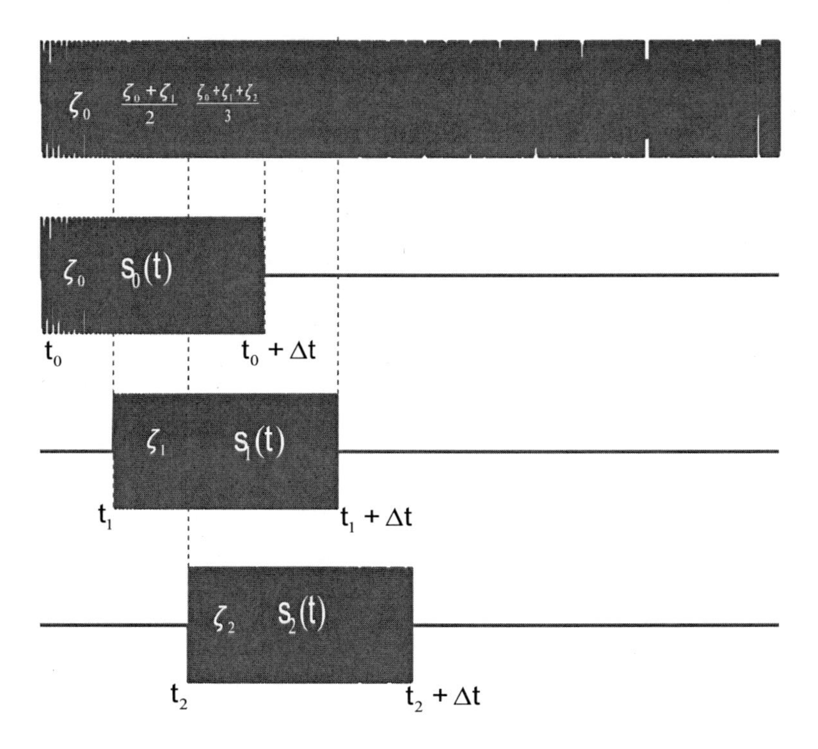

Figure 6.4.: Excitation signal segments

The following presents another problem that must be solved. When the synchronous segments are extracted from both, the whole sweep and the measured signal, then the measured signal includes the decay curve from the previous segment, which is a kind of overlap effect. In order to record the decay curve from the current sweep segment, the swept signal from the next segment is also included, as depicted in Fig. 6.5. These overlapping signals at the beginning and end of the signal will distort the impulse response.

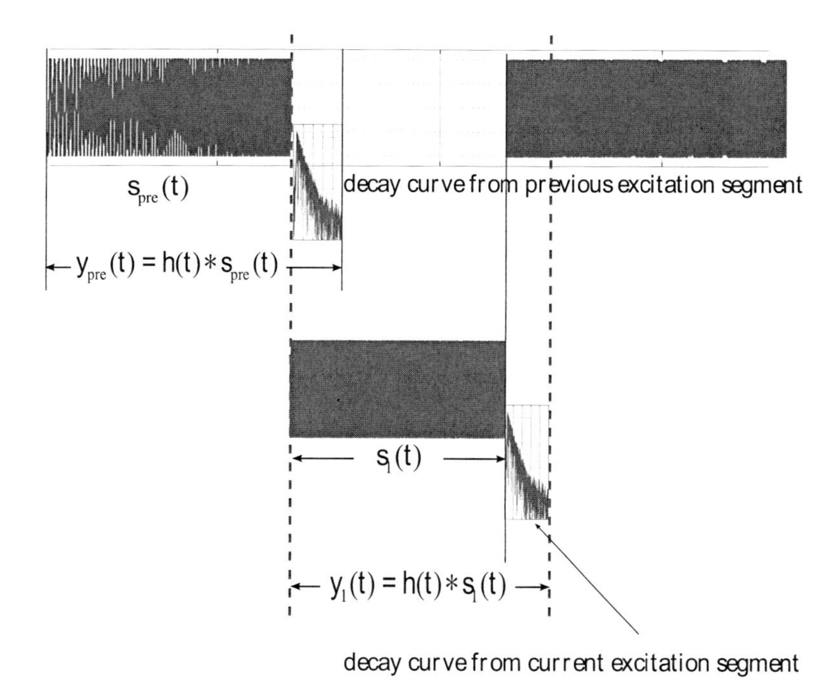

Figure 6.5.: Measured signal segments

In order to overcome this signal overlapping problem, the advantage of the sweep can be considered. Since the frequency is monotonically increasing in the sweep, only one frequency is generated at a specific instant. The decay curve from the previous sweep segment includes only the low-frequency component, and the swept signal from the following sweep segment includes only the high-frequency component, as shown in Fig. 6.6. These two additional components can be easily filtered out in the frequency domain. MLS, for example, does not have such an advantage.

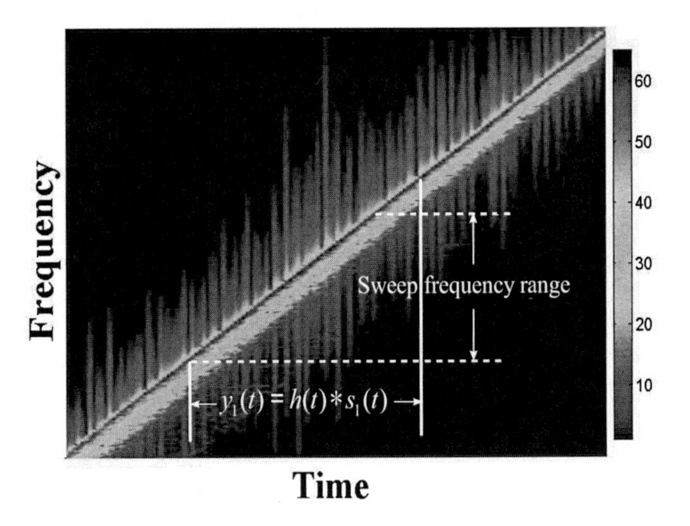

Figure 6.6.: Monotonic property of sweep

6.4. Measurement practice

When measuring the impulse response, the excitation signals must be generated through the source, as illustrated by Eq. 6.51. The measured signal also contains the convolution of the source transducer's impulse response.

$$y(t) = h(t) * h_{\text{Source transducer}}(t) * s(t) \tag{6.51}$$

Before estimating the time-stretching factor, the measured signals must first deconvolve with the impulse response of the loudspeaker; otherwise, the impulse response of the source transducer is also stretched, which might lead to unpredictable errors.

6.5. Measurement results for inter-period time variance

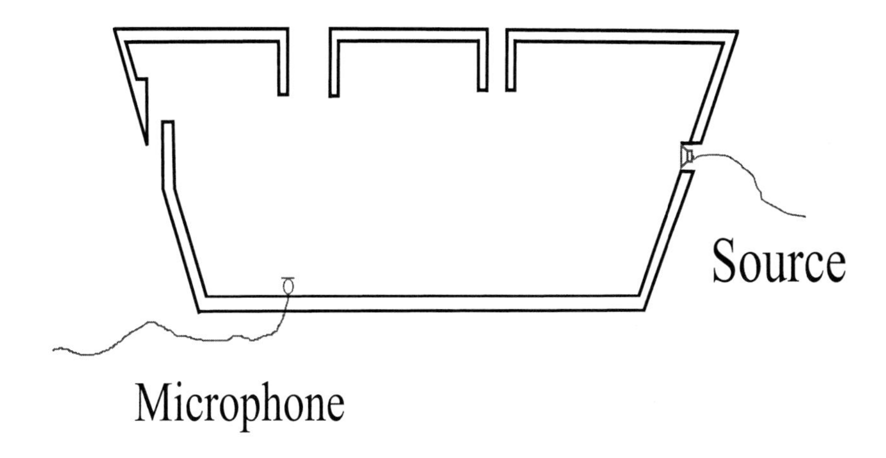

Figure 6.7.: The measurement cavity

In order to evaluate the time-stretching model under extreme conditions, a hollow steel cavity is measured (see Fig. 6.7). This could be an example for engine and gearbox components, and the air volume inside the chamber is only approximately 1.5 liters. The steel cavity is placed on a heater, and a thermometer is set inside the chamber to measure the reference temperature. During the measurement, a small loudspeaker with a diameter of 0.5 cm is placed inside this small chamber. The chamber has a large heat capacity so that the temperature distribution can be assumed to be uniform. The positions of the loudspeaker and microphone are fixed. The temperature was arbitrarily changed from 20.4°C to 28.6°C.

The temperature variance can now be observed in the impulse response in both, the time domain and the frequency domain. The impulse response at 20.4°C is measured twice in order to compare the time variance of the measurement system itself—e.g. loudspeaker and microphone's time variance. The impulse response at 24.6°C is measured once. Fig. 6.8 shows that as the temperature changes, the phase of the impulse response shifts. In Fig. 6.9, the difference between two measurements at the different temperatures along the entire time axis is compared. The green curve denotes the errors between the measurements at different temperatures $\left[h(t)_{24.6^\circ\mathrm{C}} - h(t)_{20.4^\circ\mathrm{C}}\right]$. Since large phase shifts occur, this error function is of almost the same magnitude as the reference curve.

The difference between the two measurements at the identical temperature $\left[h(t)_{\text{first measurement }20.4^\circ\text{C}} - h(t)_{\text{second measurement }20.4^\circ\text{C}}\right]$is also drawn here as a reference for the time variance of the measurement devices; this error is very small, when compared with the error caused by the temperature drift.

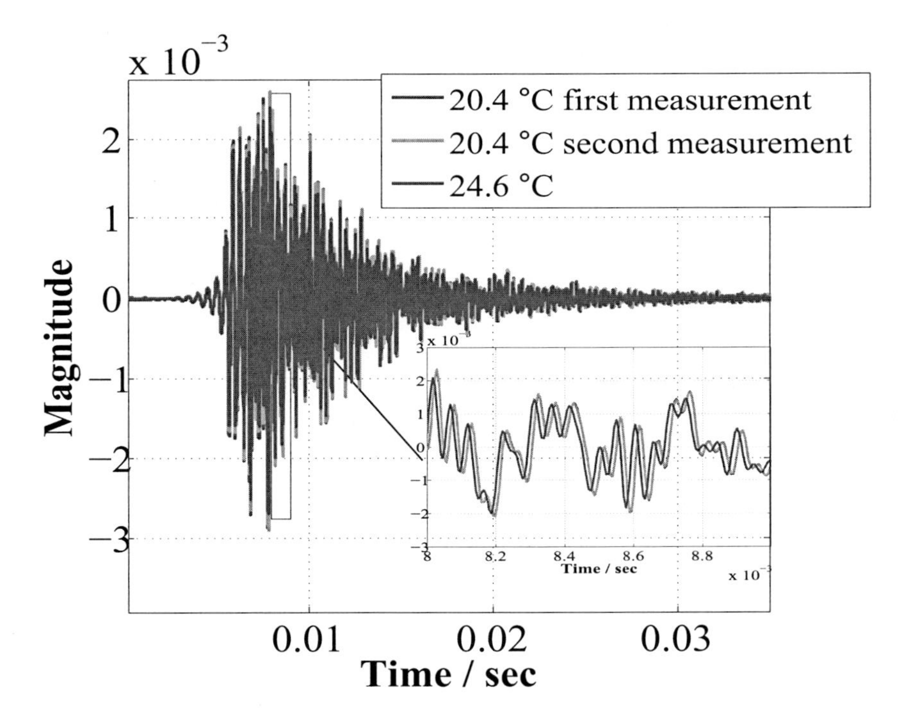

Figure 6.8.: Impulse response over various temperatures. The impulse response at 20.4°C is measured twice to ensure that no other time variance exists besides the temperature. The impulse responses at 20.4°C and 24.6°C show some phase shifts.

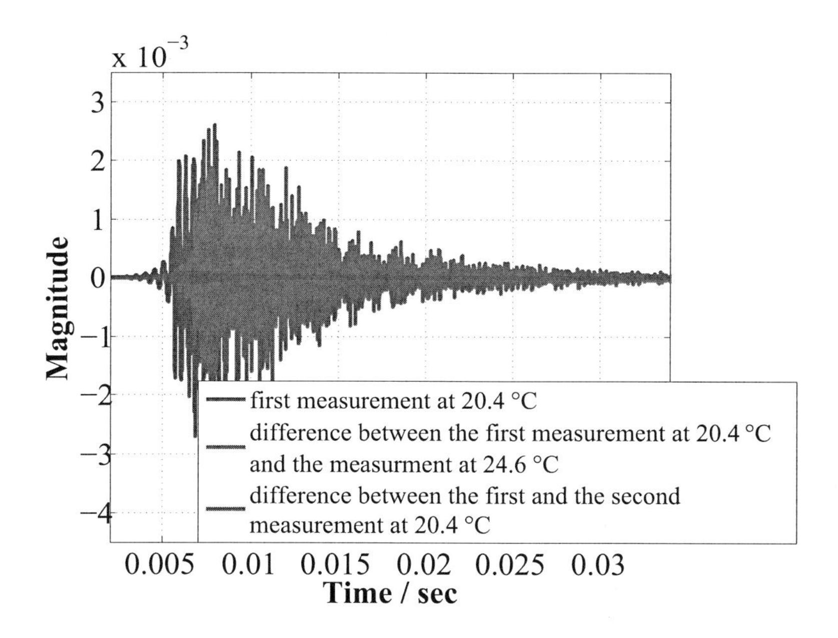

Figure 6.9.: Comparison of different measurements in the time domain

In the frequency domain (Fig. 6.10), as the temperature changes, the curve of the transfer function shifts along the frequency axis as well. The magnitude of the error function $\left[H(\omega)_{24.6^\circ\mathrm{C}} - H(\omega)_{20.4^\circ\mathrm{C}}\right]$ is as large as a single measurement. This error function $\left[H(\omega)_{24.6^\circ\mathrm{C}} - H(\omega)_{20.4^\circ\mathrm{C}}\right]$ is plotted together with the reference transfer function $H(\omega)_{20.4^\circ\mathrm{C}}$ in decibels in Fig. 6.11. The difference between the two curves denotes the relative error, which is $20\log_{10}\left|\frac{H(\omega)_{24.6^\circ\mathrm{C}} - H(\omega)_{20.4^\circ\mathrm{C}}}{H(\omega)_{20.4^\circ\mathrm{C}}}\right|$.

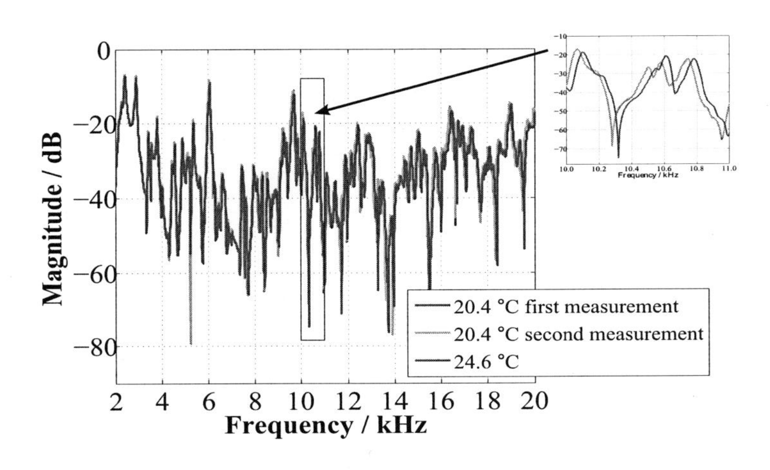

Figure 6.10.: Transfer function at various temperatures

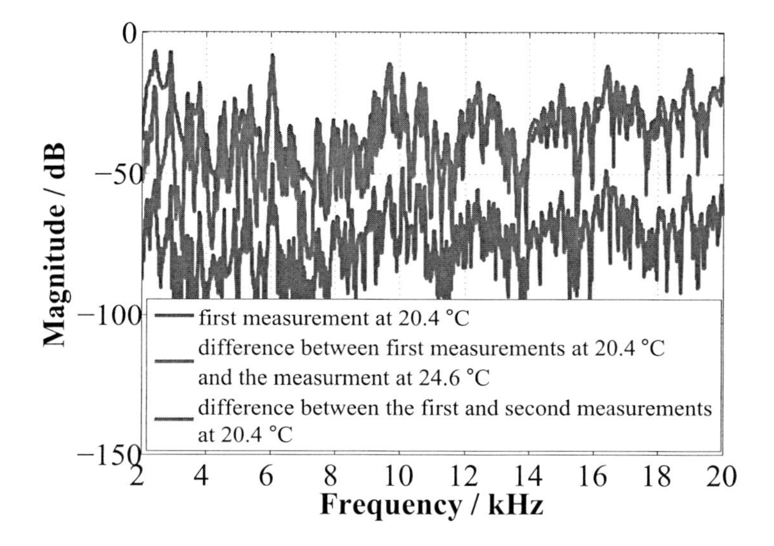

Figure 6.11.: Comparison of different measurements in the frequency domain

Fig. 6.12 and Fig. 6.13 show the performance of the time-stretching compensation. As the time-stretching factor is estimated, the curve of the impulse response at 24.6° C is stretched back to the reference impulse response 20.4° C . Compared with the errors in the cases without the time-stretching compensation—i.e. in Fig. 6.9 and Fig. 6.11—the errors are notably reduced.

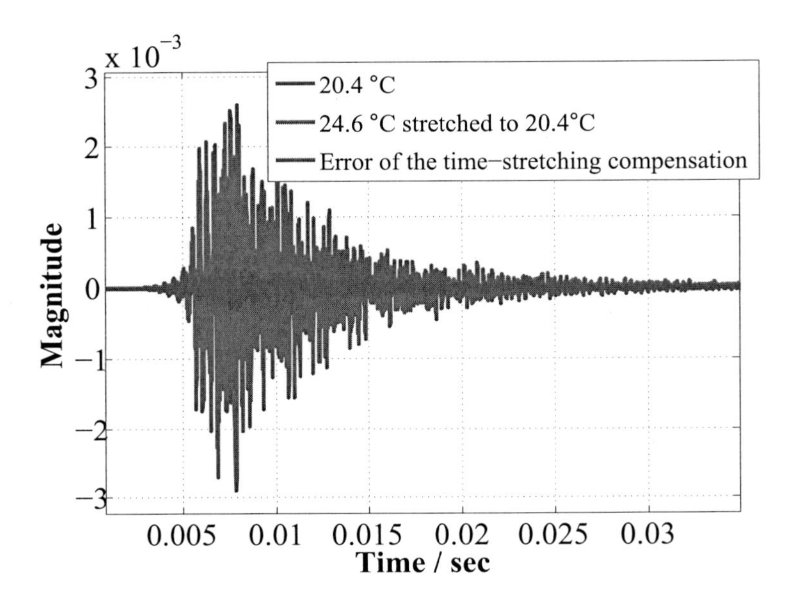

Figure 6.12.: Performance of the time-stretching compensation in the time domain

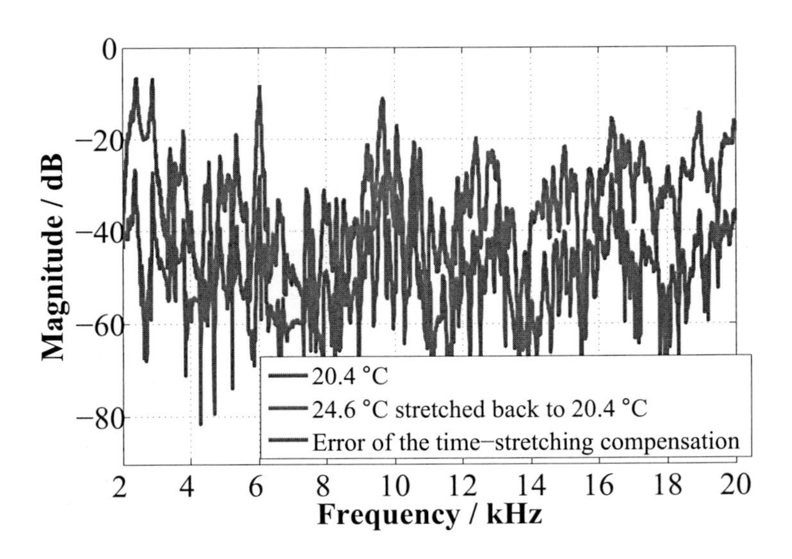

Figure 6.13.: Performance of the time-stretching compensation in the frequency domain

The time-stretching model is also used to calculate the long-time average in the temperature-variant systems. In the experiment, the temperature is changed from 20.4°C to 28.6°C, and 20 impulse responses at various temperatures are recorded. As illustrated in Fig. 6.14, directly averaging the 20 time-variant impulse responses will lead to incorrect average results, especially at high frequencies. After the impulse responses are stretched back to the reference temperature (20.4°C) before the synchronous average, the compensated average still fits very well with the transfer function at 20.4°C. The difference between the reference transfer function at 20.4°C and the averaged transfer function after the time-stretching compensation is shown in Fig. 6.15. The difference is plotted in dB,

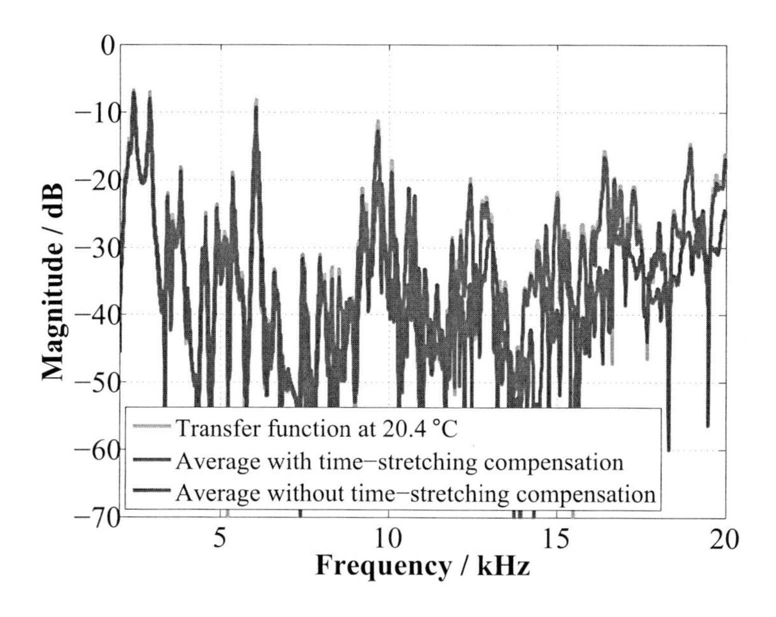

Figure 6.14.: The performance of averaging over the temperature-variant measurement with/without the time-stretching compensation

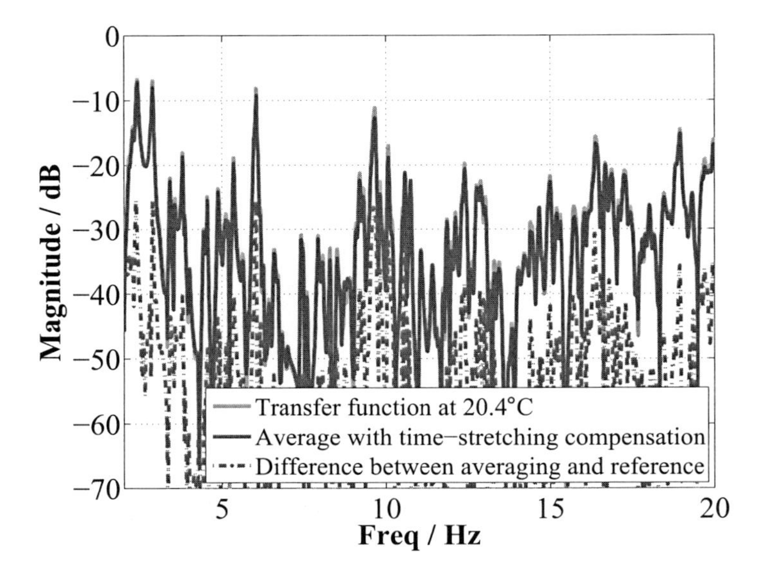

Figure 6.15.: The difference between the reference transfer function at $20.4°C$ and the averaged transfer function after the time-stretching compensation

6.6. Simulation results for intra-period time variance

The impulse response measured in Sec. 6.5 is used for intra-period time variance simulation. An 84-second-long linear sweep is generated from 2,000 Hz to 20,000 Hz. The temperature varies ±1°C during the measurement, as shown in Fig. 6.16. The purple curve is the reference transfer function at 20°C, and the green curve the simulated measured transfer function influenced by intra-period temperature variance. At the swept frequencies when the temperature drifts higher, the curve of the transfer function shifts to high frequencies, and when the temperature drifts lower, the curve of the transfer function at the corresponding frequencies shifts to low frequencies. For example, at the beginning, the temperature drifts to +0.3°C, and then increases to +1°C. During this period, the sweep is swept to low frequencies, and the curve of the transfer function shifts to a high frequency. From the 53rd to the 56th second, the temperature shift is nearly zero, and the measured transfer function is nearly equal to the reference curve between 13.4kHz and 14.2kHz. At the last part of the sweep, the temperature is lower than the reference temperature, and the curve of the measured transfer function shifts to low frequency.

The error curves in both, the time and frequency domains, are also illustrated in Fig. 6.17 and Fig. 6.18, which are defined by Eq. 6.52.

$$\begin{aligned} \text{Error time domain} &= h_{\text{Intraperiod time variance}}(t) - h_{\text{Reference}}(t) \\ \text{Error frequency domain} &= H_{\text{Intraperiod time variance}}(\omega) - H_{\text{Reference}}(\omega) \end{aligned} \tag{6.52}$$

In the frequency domain, the magnitude is plotted on the decibel scale.

The purple curve denotes the reference transfer function, which is calculated by $20\log_{10}|H_{\text{Reference}}(\omega)|$ and is overlapped by the transfer function of intra-period time variance.

The red curve denotes the difference between the reference transfer function and the time-varying transfer function. This difference is calculated by

$20\log_{10}|H_{\text{Intraperiod time variance}}(\omega) - H_{\text{Reference}}(\omega)|$.

The difference between the two curves is can be regarded as the relative error $20\log_{10}\left|\frac{H_{\text{Intraperiod time variance}}(\omega)-H_{\text{Reference}}(\omega)}{H_{\text{Reference}}(\omega)}\right|$.

Since the slight temperature drift may cause a large phase shift, the error curve is as large as the reference transfer function.

The simulated 84-second sweep and the measured signal are extracted every two seconds, and the temperature is estimated segment by segment. Since the temperature drift is assumed to be constant within this short segment, and the temperature indeed changes only slightly, a certain estimation error appears. But this small error does not matter too much. After the temperature estimation, the correct impulse response is recovered by Eq. 6.43. The results are illustrated in Fig. 6.20. After the time-warping compensation, the errors are very small—below -20 dB.

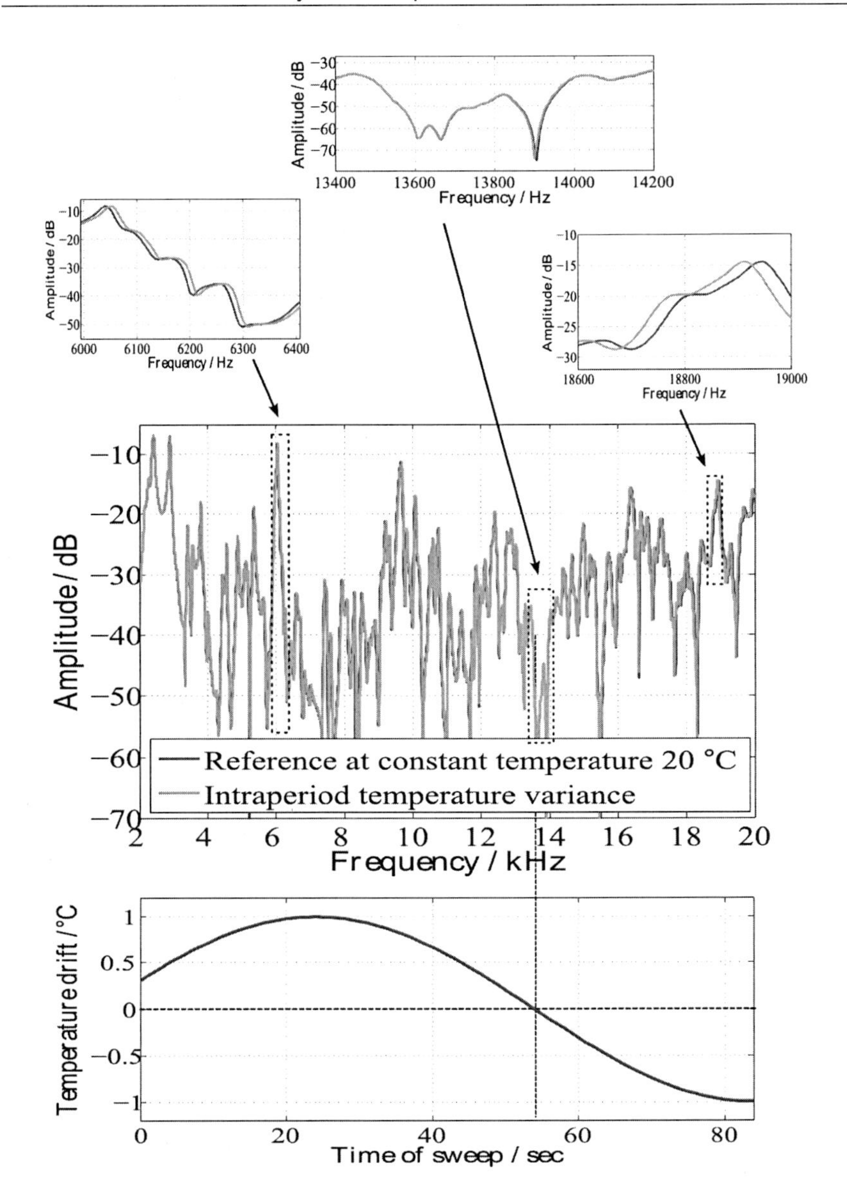

Figure 6.16.: Transfer function influenced by intra-period temperature variance

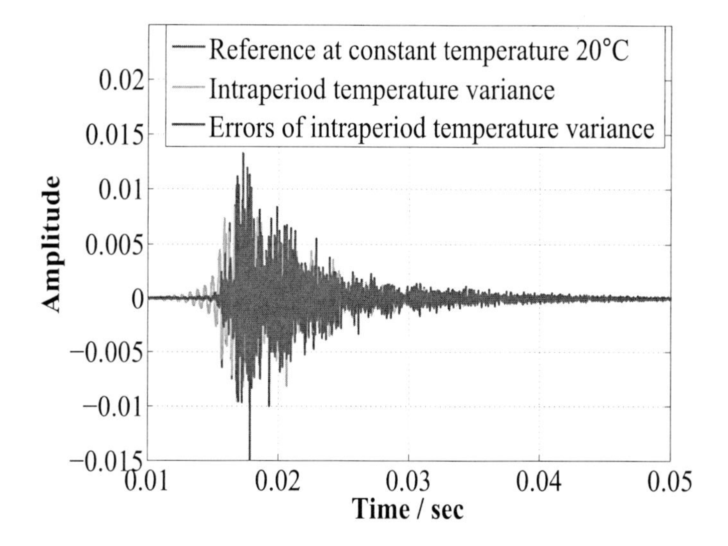

Figure 6.17.: Errors of intra-period time variance in the time domain

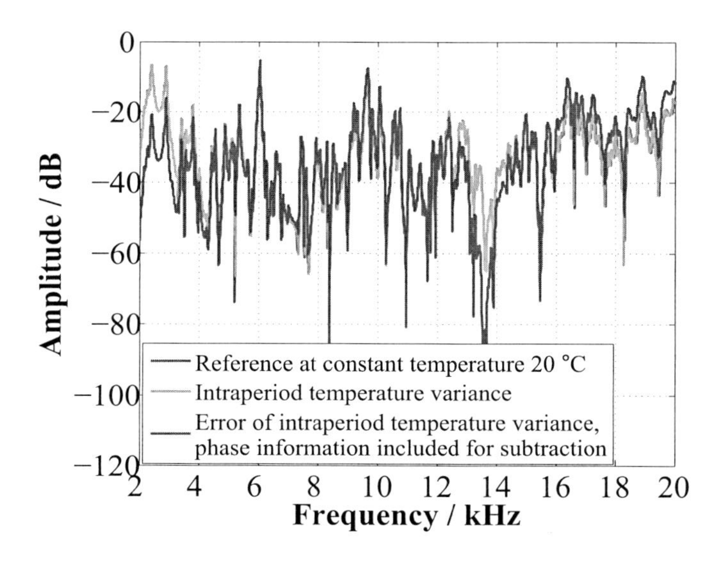

Figure 6.18.: Errors of intra-period time variance in the frequency domain

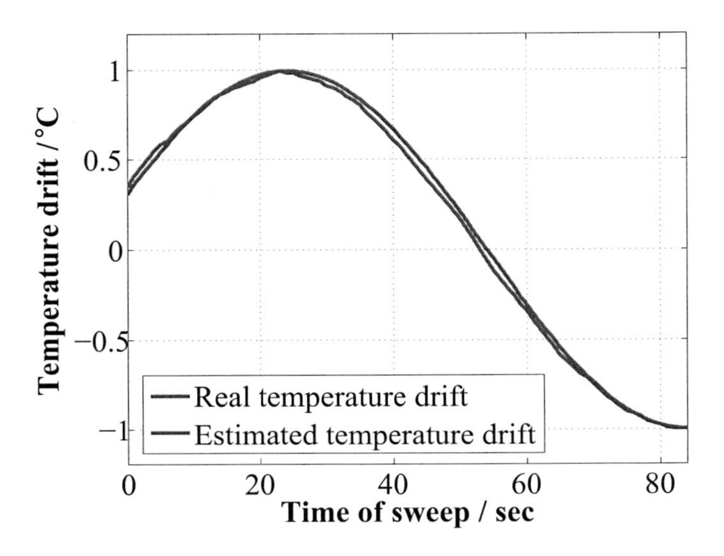

Figure 6.19.: Intra-period temperature drift estimation

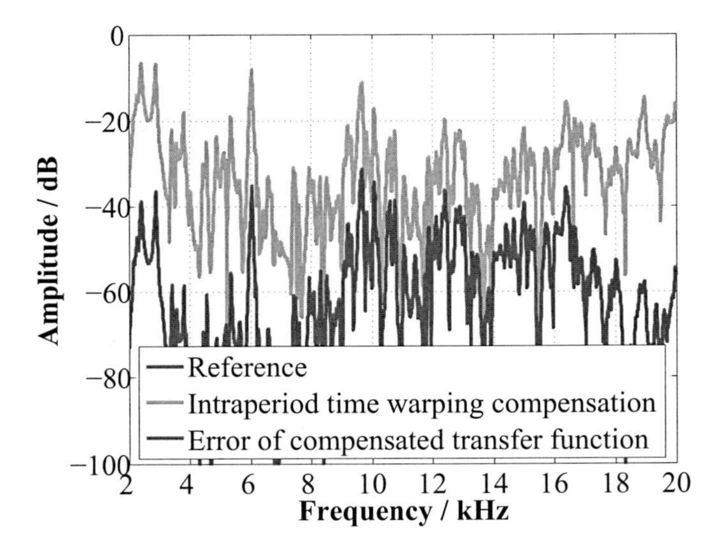

Figure 6.20.: Performance of time-warping compensation

7 Summary and Conclusion

The main motivation of this dissertation is to investigate the possibilities that reduce background noise and improve the accuracy of impulse response measurements.

Concerning the disturbing noise and aiming to deal with the disturbing noise which coincides with the excitation signal in both the time and frequency domain, the idea of separating the desired signal from the noise through statistical methods such as BSS should be discarded because the statistical source separation algorithm is an under-determined MIMO inverse problem. BSS only adds an extraneous contrast function to the parameter-estimation problem. This contrast function is the statistical independence between the source and noise, and it cannot be ensured that the correct separation is achieved when the number of parameters increases.

In problems of separating specific signal and noise and to test the robustness of multi-microphone technqiues, a standard free field method was investigated. Regarding the sound barrier measurement outdoors, instead of the subtraction technique a linear four-microphone array can be used to separate direct sound and reflected sound if sound waves impinging on each microphone are assumed almost identical. The only differences are the magnitudes and delays. In this case the parameters to be estimated are selected to be several frequency-independent magnitudes. The different magnitudes at different microphones can be estimated efficiently. However, the array size and microphone positions have to be configured delicately to avoid the problem of ill-conditioned matrix.

If the sound barrier under test is not flat and includes complicated scattering structures, the sound waves in front of the sound barrier can be considered as neither simple plane waves nor spherical waves, and the parameters cannot be assumed to be only several frequency-independent magnitudes and the delays. More parameters have to be estimated, and this microphone array fails to measure the physical impulse response.

If the system under test has a significant reverberation time and the impulse response of this system is measured in the presence of noise, a large number of parameters have to be estimated simultaneously. This estimation procedure is not possible through a single measurement because there are not enough samples. Multiple excitation signals are designed to increase the number of samples available for estimation, a large number of signal repetitions are required to ensure the independence property, and the instability of the mixing matrix inverse must be considered. In contrast, the conventional averaging approach directly estimates only the wanted parameters. It does not estimate unwanted parameters such as noise-to-receiver impulse responses, and avoids the instability problem of the matrix inverse. The averaging approach converges to the least mean square error and is more efficient and robust.

It was found that averaging is more reliable than statistical sound separation methods. But it cannot be implemented under time-variant conditions. Due to the fact that physical properties may change (in time-variant systems), further investigation were performed to check if the time variance can be modeled by a single or several latent parameters and it was investigated if averaging can be performed after compensating for the effect of time variance in advance.

In terms of the sound barrier measurement outdoors, the influence of wind must be considered. The effect of the wind can be modeled by a phase shift if the wind is a uniform flow. In more complex situations of direct and reflected waves sound in non-identical wind profiles, however, a compensation is not possible.

With regard to the impulse response of an air-borne sound measurement for a machine monitoring system, a time-warping model for inter-period and intra-period temperature variances was derived by using the Green's function theory. The effect of temperature variances can be compensated by warping temperature-dependent impulse responses to a nominal impulse response. For the inter-period time variances, the differences of two measurements is only a time-stretching factor in the impulse response. The time-stretching factor can be estimated by maximizing the cross-correlation function. The time-stretching model for the inter-period temperature variance was validated by the measurement of a model chamber.

For the intra-period temperature variance, in order to estimate the time-warping factor, the advantage of sweeps is used because the instant frequency of sweep increases monotonically with the time and the temperature shift at a specific instant influences only the corresponding frequency. The sweep is cut into short segments and the measured signal is filtered by the bandpass filter of the corresponding frequencies. The time-warping factor is estimated segments by segments by cross-correlation function. The simulation results for intra-period time variance are also illustrated.

Future research could combine the above statistical signal estimation approach and the physical model to fit specific scenarios. Regarding particular scenarios such as machine monitoring, the tested signal might have particular characteristics. If the physical and statistical properties can both be modeled correctly, the desired signal could be separated.

Solution for point source in uniform flow

$$\Delta\Psi - \frac{1}{c}\left(\frac{\partial}{\partial t} + V\frac{\partial}{\partial x}\right)^2 \Psi = 0 \tag{A.1}$$

where Ψ is the velocity potential. Separating the time factor with harmonic component $e^{i\omega t}$, the time-independent equation is [Mechel, 2008].

$$\left(\Delta - M^2\frac{\partial^2}{\partial x^2} - 2ikM\frac{\partial}{\partial x} + k^2\right)\Psi = 0 \tag{A.2}$$

It is difficult to derive the solution for a point source directly. But it can be derived in two steps.

1. A point source is moving at a constant speed along the x direction.

2. A sensor is moving at an identical speed to the source.

The equation of a moving point source is [Philip M. Morse, 1968]

$$\nabla^2\Psi - \frac{1}{c^2}\frac{\partial^2\Psi}{\partial t} = -q\left(t\right)\delta(x - Vt)\delta(y)\delta(z) \tag{A.3}$$

To solve Eq. A.3, the Lorentz transformation should be performed[Philip M. Morse, 1968].

$$\begin{bmatrix} x' \\ t' \\ y' \\ z' \end{bmatrix} = \begin{pmatrix} \gamma & -V\gamma & 0 & 0 \\ -\frac{V\gamma}{c^2} & \gamma & 0 & 0 \\ 0 & 0 & 1 & 0 \\ 0 & 0 & 0 & 1 \end{pmatrix} \begin{bmatrix} x \\ t \\ y \\ z \end{bmatrix} \tag{A.4}$$

or written as

$$\begin{bmatrix} x \\ y \\ z \\ t \end{bmatrix} = \begin{pmatrix} \gamma & V\gamma & 0 & 0 \\ \frac{V\gamma}{c^2} & \gamma & 0 & 0 \\ 0 & 0 & 1 & 0 \\ 0 & 0 & 0 & 1 \end{pmatrix} \begin{bmatrix} x' \\ t' \\ y' \\ z' \end{bmatrix} \tag{A.5}$$

performed by the Lorentz transformation

$$\nabla'^2\Psi - \frac{1}{c^2}\frac{\partial^2\Psi}{\partial t'^2} = -\gamma q(\gamma t')\delta\left(x'\right)\delta\left(y'\right)\delta\left(z'\right) \tag{A.6}$$

where $\nabla'^2 = \frac{\partial^2}{\partial x'^2} + \frac{\partial^2}{\partial y'^2} + \frac{\partial^2}{\partial z'^2}$

Eq. A.6 is equivalent to the wave equation for the point source at the fixed position.

Hence, the solution of Eq. A.6 is

$$\Psi\left(r', t'\right) = \gamma\frac{q\left[\gamma\left(t' - \frac{r'}{c}\right)\right]}{4\pi r'} \tag{A.7}$$

Transforming back to the (x, y, z, t) coordinate systems,

$$\Psi\left(r, t\right) = \gamma\frac{q\left[\gamma\left((t - \frac{Vx}{c^2})\gamma - \frac{\sqrt{\left[(x-Vt)\gamma\right]^2 + y^2 + z^2}}{c}\right)\right]}{4\pi\sqrt{\left[(x - Vt)\gamma\right]^2 + y^2 + z^2}} \tag{A.8}$$

This is the solution for a moving point source. On moving the sensor at the same speed as that of the source, the effect is equivalent to doing a Galileo transformation.

$$\begin{aligned} x'' &= x - Vt \\ y'' &= y \\ z'' &= z \\ t'' &= t \end{aligned} \tag{A.9}$$

$$\Psi\left(r'', t''\right) = \gamma \frac{q\left(t'' + \frac{V\gamma^2 x''}{c^2} - \gamma \frac{\sqrt{\gamma^2 x''^2 + y''^2 + z''^2}}{c}\right)}{4\pi\sqrt{\gamma^2 x''^2 + y''^2 + z''^2}} \tag{A.10}$$

for the harmonic waves $q(t) = e^{i\omega t}$

$$\Psi\left(r'', t''\right) = \gamma \frac{e^{i\omega\left(t'' + \frac{V\gamma^2 x''}{c^2} - \gamma \frac{\sqrt{\gamma^2 x''^2 + y''^2 + z''^2}}{c}\right)}}{4\pi\sqrt{\gamma^2 x''^2 + y''^2 + z''^2}} \tag{A.11}$$

One can easily prove that the solution Eq. A.11 satisfies the convective wave equations Eq. A.1 and Eq. A.2.

Bibliography

ASTM E1050 standard test method for impedance and absorption of acoustical materials using a tube, two microphones and a digital frequency analysis system.

1998. EN ISO 10534-2:1998 acoustics – determination of sound absorption coefficient and impedance in impedance tubes – part 2: Transfer-function method.

2003. ISO 354:2003 acoustics – measurement of sound absorption in a reverberation room.

Abhayapala, TD; Kennedy, R. W. R. 2000. Nearfield broadband array design using a radially invariant modal expansion. *Journal of The Acoustical Society of America*, 107(1):392–403.

Adali, Tülayi; Li, H. N. M. C. J.-F. F. 2008. Complex ica using nonlinear functions. *IEEE Transactions on Signal Processing*, 56(9):4536–4544.

Alan Victor Oppenheim, Alan S. Willsky, w. S. H. N. 1997. *Signals and Systems.* Prentice-Hall International.

Barkat, M. 2005. *Signal Detection and Estimation.* Artech House.

Belouchrani, A; AbedMeraim, K. C. J. M. E. 1997. A blind source separation technique using second-order statistics. *IEEE Transactions on Signal Processing*, 45(2):434–444.

Berman, J. Michael; Fincham, L. R. 1977. The application of digital techniques to the measurement of loudspeakers. *Journal of Audio Engineering Society*, 25(6):370–384.

Bingham, E; Hyvarinen, A. 2000. A fast fixed-point algorithm for independent component analysis of complex valued signals. *International journal of neural systems*, 10(1):1–8.

Borish, J. 1985. Self-contained crosscorrelation program for maximum-length sequences. *Journal of the Audio Engineering Society*, 888-891(11):11.

Borish, Jeffrey; Angell, J. B. 1983. An efficient algorithm for measuring the impulse response using pseudorandom noise. *Journal of the Audio Engineering Society*, 31(7/8):478–488.

Buchner, H; Aichner, R. K. W. 2005. A generalization of blind source separation algorithms for convolutive mixtures based on second-order statistics. *IEEE Transactions on Speech and Audio Processing*, 13(11):120–134.

Capon, J. 1969. High-resolution frequency-wavenumber spectrum analysis. *Proceedings of the IEEE*, 57(8):1408–1418.

Cardoso, JF; Souloumiac, A. 1993. Blind beamforming for non-gaussian signals. *IEE Proceedings-F Radar and Signal Processing*, 140(6):362–370.

Chou, T. 1995a. Frequency-independent beamformer with low response error. *International Conference on Acoustics, Speech, and Signal Processing, ICASSP-95*, 5:2995 – 2998.

Chou, T. C. 1995b. *Broadband Frequency-independent Beamforming.* PhD thesis, Massachusetts Institute of Technology.

Cobo, P; Fernandez, A. C. M. 2007. Measuring short impulse responses with inverse filtered maximum-length sequences. *Applied Acoustics*, 68(7):820–830.

Comon, P; Rota, L. 2003. Blind separation of independent sources from convolutive mixtures. *IEICE Trans. on Fundamentals of Elec. Com. Comput. Sciences, , March 2003*, E86-A(3).

Comon, P. 1994. Independent component analysis, a new concept. *Signal Processing*, 36(3):287–314.

Comon, P. 1996. Contrasts for multichannel blind deconvolution. *IEEE Signal Processing Letters*, 3(7):209–211.

Darren B. Ward, R. A. K. and Williamson, R. C. 1995. Theory and design of broadband sensor arrays with frequency invariant far-field beam patterns. *Journal of the Acoustical Society of America*, 97(2):1023–1034.

De Boor, C. 2001. *A practical guide to splines*, volume 27. Springer Verlag.

Douglas, S. C. 2007. Fixed-point algorithms for the blind separation of arbitrary complex-valued non-gaussian signal mixtures. *EURASIP Journal on Applied Signal Processing*, SI(36525):83–83.

Dudgeon, D. H. J. E. 1993. *Array Signal Processing: Concepts and Techniques.* Prentice Hall.

E. C. Wente, E. H. Bedell, K. D. S. J. 1935. A high speed level recorder for acoustic measurements. *Journal of the Acoustical Society of America*, 6(3):121–129.

E. Ward Cheney, D. R. K. 2007. *Numerical Mathematics and Computing.* Cengage Learning.

Eriksson, J. 2004. Complex-valued ica using second order statistics. In *Machine Learning for Signal Processing, 2004. Proceedings of the 2004 14th IEEE Signal Processing Society Workshop.*

Farina, A. 2000. Simultaneous measurement of impulse response and distortion with a swept-sine technique. In *108th AES Convention.*

Farina, A. 2007a. Advancements in impulse response measurements by sine sweeps. In *122nd AES Convention.*

Farina, A. 2007b. Impulse Response Measurements. In *23RD NORDIC SOUND SYMPOSIUM.*

Franklin A. Graybill, H. K. I. 1994. *Regression Analysis: Concepts and Applications.* Duxbury Press.

Fumiaki Satoh, Mitsuru Nagayama, H. T. 2002. Influence of time-variance in auditorium on impulse response measurement. In *Proceedings of Forum Acusticum Sevilla.*

Fumiaki Satoh, Jin Hirano, S. S. H. T. 2004. Comparison between the mls and tsp methods for room impulse response measurement under time-varying condition. In *Proceedings of International Symposium on Room Acoustics : Design and Science.*

Garai, M. 1993. Measurement of the sound-absorption coefficient in situ: The reflection method using periodic pseudo-random sequences of maximum length. *Applied Acoustics*, 39(1-2):119–139.

Garai, M. 2011. Noise reducing devices acting on airborne sound propagation test method for determining the acoustic performance intrinsic characteristics in situ values of airborne sound insulation under direct sound field conditions. (http://www.quiesst.eu). Technical report, QUIESST D3.3.

Hall, C. A. and Meyer, W. 1976. Optimal error bounds for cubic spline interpolation. *Journal of Approximation Theory*, 16(2):105 – 122.

Haykin, S., editor 1994. *Blind Deconvolution.* Prentice Hall.

Heyser, R. C. 1967. Acoustical measurements by time delay spectrometry. *Journal of the Audio Engineering Society*, 15(4):370–382.

Heyser, R. C. 1969a. Loudspeaker phase characteristics and time delay distortion: Part 1. *Journal of The Audio Engineering Society*, 17(1):30–41.

Heyser, R. C. 1969b. Loudspeaker phase characteristics and time delay distortion: Part 2. *Journal of The Audio Engineering Society*, 17:130–137.

Hyvarinen, Aapo; Karhunen, J. E. O. 2001. *Independent component analysis.* John Wiley & Sons.

Ishimaru, A. 1962. Theory of unequally-spaced arrays. *Institute of Radio Engineers Transactions on Antennas and Propagation*, AP-10(6):691–702.

Ishimaru, A; Chen, Y. 1965. Thinning and broadbanding antenna arrays by unequal spacings. *IEEE Transactions on Antennas and Propagation*, AP13(1):34 – 42.

Jutten, P. C., editor 2010. *Handbook of Blind Source Separation: Independent Component Analysis and Applications.* Amsterdam ; Boston : Elsevier.

Kellermann, H. B. R. A. W. 2004. *Audio Signal Processing: For Next-Generation Multimedia Communication Systems*, chapter 10, Blind Source Separation for Convolutive Mixtures: A Unified Treatment, pages 255–293. Springer.

Kellermann, H. B. W. 2010. *Speech Dereverberation*, chapter 10 TRINICON for Dereverberation of Speech and Audio Signals, pages 311–385. Springer.

Kennedy, RA; Ward, D. A. T. 1999. Nearfield beamforming using radial reciprocity. *IEEE Transactions on Signal Processing*, 47(1):33–40.

Kennedy, RA; Abhayapala, T. W. D. 1998. Broadband nearfield beamforming using a radial beampattern transformation. *IEEE Transactions on Signal Processing*, 46(8):2147–2156.

Kuttruff, H. 2007. *Acoustics: an introduction.* Taylor & Francis.

Lacoume, J.J.; Ruiz, P. 1988. Sources indentification: a solution based on the cumulants. In *Fourth Annual ASSP Workshop on Spectrum Estimation and Modeling*, pages 199–203.

Lanoye, R; Vermeir, G. L. W. K. R. M. V. 2006. Measuring the free field acoustic impedance and absorption coefficient of sound absorbing materials with a combined particle velocity-pressure sensor. *Journal of the Acoustical Society of America*, 119(5):2826–2831.

Lawrence E. Kinsler, Austin R. Frey, A. B. C. J. V. S. 1982. *Fundamental of Acoustic.* John Wiley & Sons.

Maurice George Kendall, A. S. 1961. *The Advanced Theory of Statistics: Volume 2: Inference and Relationship.* Hafner Publishing Company.

Mechel, F., editor 2008. *Formulas of Acoustics.* Springer-Verlag Berlin Heidelberg New York.

Mikio Tohyama, T. K. 1998. *Fundamentals of Acoustic Signal Processing.* Academic Press; 1 edition.

Mommertz, E. 1995. Angle-dependent in-situmeasurements of reflectioncoefficients using asubtractiontechnique. *Applied Acoustics*, 46(3):251–263.

Moreau, E; Pesquet, J. 1997. Generalized contrasts for multichannel blind deconvolution of linear systems. *IEEE SIGNAL PROCESSING LETTERS*, 4(6):182–183.

Mulholland, J. D. K. 1979. An impulse method of measuring normal impedance at oblique incidence. *Journal of Sound and Vibration*, 67(1):135–149.

Müller, S. and Massarani, P. 2001. Transfer-function measurement with sweeps. *Journal of the Audio Engineering Society*, 49(6):443–471.

Neeser, F. D. 1993. Proper complex random processes with applications to information theory. *IEEE Transactions on Information Theory*, 39(4):1293–1302.

Nocke, C. 2000. In-situ acoustic impedance measurement using a free-field transfer function method. *Applied Acoustics*, 59(3):253–264.

Parra, L. 2006. Steerable frequency-invariant beamforming for arbitrary arrays. *Journal of the Acoustical Society of America*, 119(6):3839–3847.

Philip M. Morse, K. U. I. 1968. *Theoretical Acoustics.* McGraw-Hill.

Rife, Douglas D.; Vanderkooy, J. 1989. Transfer-function measurement with maximum-length sequences. *Journal of the Audio Engineering Society*, 37(6):419–444.

Ryan, JG; Goubran, R. 2000. Array optimization applied in the near field of a microphone array. *Ieee Transactions on Speech and Audio Processing*, 8(2):173–176.

Satoh, F.; Hirano, J. S. S. T. H. 2004. Sound insulation measurement using a long swept-sine signal. In *8th International Congress on Acoustics.*

Schreier, Peter J.; Scharf, L. L. 2010. *Statistical Signal Processing of Complex-Valued Data: The Theory of Improper and Noncircular Signals.* Cambridge University Press; 1 edition.

Schroeder, M. R. 1979. Integrated impulse method measuring sound decay without using impulses. *Journal of the Acoustical Society of America*, 66(2):497–500.

Snyman, J. A. 2005. *Practical Mathematical Optimization: An Introduction to Basic Optimization Theory and Classical and New Gradient-Based Algorithms.* Springer.

Strang, G. 1988. *Linear Algebra and Its Applications.* Harcourth Brace Jovanovich; 3 edition.

Svensson, U. and Nielsen, J. 1999. Errors in MLS measurements caused by time variance in acoustic systems. *Journal of the Audio Engineering Society*, 47(11):907–927. 100th Convention of the Audio-Engineering-Society, COPENHAGEN, DENMARK, MAY 11-14, 1996.

Tohyama, M. 2011. *Sound and Signals.* Springer;.

Trees, H. L. V. 2002. *Optimum Array Processing: Part IV of Detection, Estimation, and Modulation Theory.* Wiley-Interscience.

Vorländer, M. and Kob, M. 1997. Practical aspects of MLS measurements in building acoustics. *Applied Acoustics*, 52(3-4):239–258. Meeting of the Building Acoustics Working Group of the CIB, ANTWERP, BELGIUM, APR, 1996.

Wehr, R; Haider, M. . C. M. G. S. B. S. 2013. Measuring the sound absorption properties of noise barriers with inverse filtered maximum length sequences. *Applied Acoustics*, 74(5):631–639.

Weisberg, S. 2005. *Applied Linear Regression.* John Wiley & Sons, Inc.

Xiang, P. R. N. 2010. On the subtraction method for in-situ reflection and diffusion coefficient measurements. *Journal of the Acoustical Society of America*, 127(3):EL99–EL104.

Yuzawa, M. 1975. A method of obtaining the oblique incident sound absorption coefficient through an on-the-spot measurement. *Applied Acoustics*, 8:27–41.

Acknowledgments

I would never have been able to finish my dissertation without the guidance of my supervisor, help from my collegues and support from my families and friends.

Firstly, I would like to appreciate Prof. Dr. rer. nat. Michael Vorländer , head of the Institute of Technical Acoustics (ITA) of the RWTH Aachen University, for accepting me and giving me the opportunity to peruse my PHD in this renowned institute. Thank you for your excellent guidance and patience to show me the way to do the scientific research.

I also appreciate Prof. Dr.-Ing. Michael Möser from Technische Universität Berlin(TU Berlin) for taking on the role as the second examiner of my thesis. I must further thank Dr.-Ing. Gottfried Behler and Prof. Dr.-Ing. Janina Fels for the patience to answer my academic and non-academic questions.

I really appreciate to all my colleagues. Bruno Masiero always actively not only helped me on the academic problems but also taught me on starting a new life in Germany. Elena Shabalina encouraged me quite a lot when I was not used to my study at the beginning. Pascal Dietrich gave me the most important inspiration which leads to the main contribution of my dissertation. Martin Pollow, Markus Müller-Trapet and Martin Guski, I cannot tell a specific event on what you have helped me and although you didn't help me every day, but you have indeed given me a hand every week, and I benefited from your help and learned a lot. Rob Opdam, I think I got massive inspirations from you, if you came to ITA one year earlier, I could have more outcomes. Ingo Witew, I think the most important thing I learned from you is "the way of thinking". Now I'm teaching my students "the way of thinking" which I learned from you.

Thanks to all other colleagues Jan Köhler, Dr. Marc Aretz, Dr. Matthias Lievens, Ramona Bomhardt, Dr. Roman Scharrer, Dr. Dirk Schröder, Dr. Sebastian

Fingerhuth, Frank Wefers, Sönke Pelzer, Josefa Oberem, Johannes Klein, Renzo Vitale for the friendly research atmosphere.

Thanks to Rolf Kaldenbach and Uwe Schlömer for the professional technical support. Thanks to the institute's secretary, Karin Charlier for helping me to prepare all my documents for the doctoral defense.

I'm obliged to thank the China Scholarship Council for the financial support to do my research in Germany.

I'm also grateful to Prof. Jun Yang and Prof. Xiaobin Chen at Institute of Acoustics, Chinese Academy of Science, who offer me a job and give me time to revise my dissertation in their Institute.

Finally, I have to thank my parents, Huali Xiang and Dianchao Wang for always supporting me.

Curriculum Vitae

Personal Data

	Xun Wang
21. 03. 1983	born in Zigong, Sichuan, China

Education

9.1989-7.1995	Elementary School, Zigong, Sichuan, China
9.1995-7.1998	No.28 Junior School, Zigong, Sichuan , China
9.1998-7.2001	Shuguang High School, Zigong, Sichuan, China

Higher Education

9.2001-7.2005	Bachelor in Applied Physics Tianjin University, Tianjin, China
9.2005-7.2008	Master in Condensed Matter Physics Peking University, Beijing, China

Professional Experience

10.2008-12.2012	Research assistant Institute of Technical Acoustics (ITA) RWTH Aachen University, Germany
12.2012-12.2013	Research assistant Institute of Acoustics Chinese Academy of Science, Beijing, China

Aachen, Germany, January 10, 2014

Bisher erschienene Bände der Reihe

Aachener Beiträge zur Technischen Akustik

ISSN 1866-3052

1	Malte Kob	Physical Modeling of the Singing Voice ISBN 978-3-89722-997-6 40.50 EUR
2	Martin Klemenz	Die Geräuschqualität bei der Anfahrt elektrischer Schienenfahrzeuge ISBN 978-3-8325-0852-4 40.50 EUR
3	Rainer Thaden	Auralisation in Building Acoustics ISBN 978-3-8325-0895-1 40.50 EUR
4	Michael Makarski	Tools for the Professional Development of Horn Loudspeakers ISBN 978-3-8325-1280-4 40.50 EUR
5	Janina Fels	From Children to Adults: How Binaural Cues and Ear Canal Impedances Grow ISBN 978-3-8325-1855-4 40.50 EUR
6	Tobias Lentz	Binaural Technology for Virtual Reality ISBN 978-3-8325-1935-3 40.50 EUR
7	Christoph Kling	Investigations into damping in building acoustics by use of downscaled models ISBN 978-3-8325-1985-8 37.00 EUR
8	Joao Henrique Diniz Guimaraes	Modelling the dynamic interactions of rolling bearings ISBN 978-3-8325-2010-6 36.50 EUR
9	Andreas Franck	Finite-Elemente-Methoden, Lösungsalgorithmen und Werkzeuge für die akustische Simulationstechnik ISBN 978-3-8325-2313-8 35.50 EUR

10	Sebastian Fingerhuth	Tonalness and consonance of technical sounds ISBN 978-3-8325-2536-1 42.00 EUR
11	Dirk Schröder	Physically Based Real-Time Auralization of Interactive Virtual Environments ISBN 978-3-8325-2458-6 35.00 EUR
12	Marc Aretz	Combined Wave And Ray Based Room Acoustic Simulations Of Small Rooms ISBN 978-3-8325-3242-0 37.00 EUR
13	Bruno Sanches Masiero	Individualized Binaural Technology. Measurement, Equalization and Subjective Evaluation ISBN 978-3-8325-3274-1 36.50 EUR
14	Roman Scharrer	Acoustic Field Analysis in Small Microphone Arrays ISBN 978-3-8325-3453-0 35.00 EUR
15	Matthias Lievens	Structure-borne Sound Sources in Buildings ISBN 978-3-8325-3464-6 33.00 EUR
16	Pascal Dietrich	Uncertainties in Acoustical Transfer Functions. Modeling, Measurement and Derivation of Parameters for Airborne and Structure-borne Sound ISBN 978-3-8325-3551-3 37.00 EUR
17	Elena Shabalina	The Propagation of Low Frequency Sound through an Audience ISBN 978-3-8325-3608-4 37.50 EUR
18	Xun Wang	Model Based Signal Enhancement for Impulse Response Measurement ISBN 978-3-8325-3630-5 34.50 EUR